ADAPTATION AND NATURAL SELECTION
A Critique of Some Current Evolutionary Thought

ADAPTATION AND NATURAL SELECTION

A Critique of Some Current Evolutionary Thought

GEORGE C. WILLIAMS

With a new foreword by Richard Dawkins

Princeton University Press
Princeton and Oxford

Copyright © 1966 by Princeton University Press

Copyright © renewed 1992

Preface to the first Princeton Science Library edition
copyright © 1966 by Princeton University Press

Foreword to the new Princeton Science Library edition
copyright © 2019 by Richard Dawkins

Published by Princeton University Press,
41 William Street, Princeton, New Jersey 08540

In the United Kingdom: Princeton University Press,
6 Oxford Street, Woodstock, Oxfordshire OX20 1TR

press.princeton.edu

Cover design by Michael Boland for
the bolanddesignco.com

Cover art: iStock

All Rights Reserved

New paperback ISBN 978-0-691-18286-5

Library of Congress Control Number: 2018947920

British Library Cataloging-in-Publication Data is available

This book has been composed in New Caledonia LT Std

Printed on acid-free paper. ∞

Printed and bound by CPI Group (UK) Ltd, Croydon, CR0 4YY

10 9 8 7 6 5 4 3 2 1

Contents

FOREWORD ix

PREFACE (1996) xvii

PREFACE xxiii

1. INTRODUCTION 3

Evolutionary adaptation is a special and onerous concept that should not be used unnecessarily, and an effect should not be called a function unless it is clearly produced by design and not by chance. When recognized, adaptation should be attributed to no higher a level of organization than is demanded by the evidence. Natural selection is the only acceptable explanation for the genesis and maintenance of adaptation.

2. NATURAL SELECTION, ADAPTATION, AND
   PROGRESS 20

Natural selection can be effective only where there are certain quantitative relationships among sampling errors, selection coefficients, and rates of random change. The selection of alternative alleles in Mendelian populations meets the requirements. Other conceivable kinds of selection do not. Selection of alternative alleles works only with an immediate better-vs.-worse among individuals in a population, and the question of population survival is irrelevant. Once a certain level of complexity is evolved, selection will maintain adaptation by occasionally substituting one adaptive character for another, but this will not result in any of the kinds of cumulative progress that have been envisioned.

CONTENTS

3. NATURAL SELECTION, ECOLOGY, AND
   MORPHOGENESIS   56

The gene is selected through a complex interaction with its environment, which can usefully be considered to include several levels: the genetic, the somatic, and the ecological. The ecological has many aspects, one of which, the "demographic," is given special treatment. Age-specific birth rates and death rates are important factors in the selection of developmental rates and other aspects of life cycles. The importance of genetic assimilation as a creative factor is minimized.

4. GROUP SELECTION   92

Selection at the genic level can produce adaptive organization of individuals and family groups. Any adaptive organization of a population must be attributed to the selection of alternative populations. Reasons are advanced for doubting, *a priori*, the effectiveness of such group selection. *Organic adaptations*, which function to maximize the genetic survival of individuals, are distinguished from *biotic adaptations*, which would be designed to perpetuate a population or more inclusive group.

5. ADAPTATIONS OF THE GENETIC SYSTEM   125

Phenomena relating to the genetic system, such as dominance, diploidy, sex-determining mechanisms, and the distribution of sexual and asexual reproduction in life cycles are easily explained as short-term organic adaptations. The survival and evolution of groups are fortuitous consequences of these adaptations and of their occasional malfunctioning, as in mutation and introgression. There is no respectable evidence of mechanisms for maintaining evolutionary plasticity or of any other biotic adaptations of the genetic system.

CONTENTS

6. REPRODUCTIVE PHYSIOLOGY AND BEHAVIOR   158

Variations in the intensity of reproductive effort, and in the manner in which it is expended, seem designed to maximize the reproductive success of the reproducing individuals. Attention is given to the evolution of fecundity, viviparity, gregarious reproduction, and differences between the sexes in reproductive behavior. These phenomena support the conclusion that the goal of an individual's reproduction is not to perpetuate the population or species, but to maximize the representation of its own germ plasm, relative to that of others in the same population.

7. SOCIAL ADAPTATIONS   193

Selection within a population can lead to cooperative relations among closely related individuals, because the benefits of cooperation would go mainly to individuals with the genetic basis of cooperation, rather than to those of alternative genetic makeup. Selection at the genic level thus explains insect societies and analogous developments in other organisms. Other apparent examples of altruism are explained as misplaced parental behavior. They represent imperfections in the mechanisms that normally regulate the timing and execution of parental behavior. Benefits to groups often arise as incidental statistical consequences of individual activities, just as harmful effects may accumulate in the same way.

8. OTHER SUPPOSEDLY GROUP-RELATED ADAPTATIONS   221

Various supposed biotic adaptations, such as poisonous flesh, senescence, and genetically heterogeneous somata, are examined and found to be spurious or inconclusive. The regulation of population size is shown to arise from individual adaptations or purely physical principles rather than as an adaptive organization of the group.

Similar arguments are used against the concept of an adaptive organization of ecological communities or more inclusive entities.

9. THE SCIENTIFIC STUDY OF ADAPTATION   251

For a given biological mechanism there are no established principles and procedures for answering the question, "What is its function?" Objectively determined answers to such questions would facilitate progress in many fields of biology, but they must await development of special concepts for the study of adaptation as a general principle. Teleonomy is a suitable name for this special field of study.

LITERATURE CITED   275

INDEX   291

Foreword

THE NEO-DARWINIAN SYNTHESIS of the 1930s and 40s was a collective Anglo-American achievement, defined by a recognizable "canon" of seminal books, those of Fisher, Haldane, Mayr, Dobzhansky, Simpson and others. Julian Huxley's *Evolution: the Modern Synthesis* bequeathed its title to the whole movement although for its theoretical content it doesn't stand out. If I were asked to nominate one book from the second half of the twentieth century that deserves to take an honored place alongside the "canon" of the 30s and 40s, I would choose *Adaptation and Natural Selection* by George C. Williams. On opening it I have the feeling of being ushered into the presence of a penetrating and outstanding mind, the same feeling I get, indeed, from reading *The Genetical Theory of Natural Selection*, although Williams, unlike Fisher, was no mathematician. In George Williams we have an author of immense learning and incisive critical intelligence, who thought deeply about every aspect of evolution and ecology. Williams not only enlarged the Synthesis, he exposed with great clarity where many of its followers had gone astray, even in some cases the original authors themselves. This is a book that every serious student of biology must read, a book that irrevocably changes the way we look at life. Throughout my career as an Oxford tutor, I obviously recommended many books to my students. But I think this was the only one I insisted that all should read.

Here's a list of major mistakes a student is likely to make before reading this book, but will not make afterwards.

FOREWORD

"Mutations are an adaptation to speed up evolution." "Dominance hierarchies are an adaptation to make sure the strongest individuals reproduce." "Territoriality is an adaptation to space the species out and beneficially limit the population." "Sex ratios are optimized to make the best use of species resources." "Death from old age is an adaptation to clear superannuated individuals out of the way and make room for the young." "Natural selection favors species that resist extinction." "Species parcel out niches for the benefit of a balanced ecosystem." "Predators hunt 'prudently,' taking care not to deplete prey that they are going to need in the future." "Individuals limit their reproduction to avoid overpopulation."

Adaptation is the first word in the title and the book is largely a plea for a proper, scientific study of adaptation—a scientific *teleonomy*, to adopt Pittendrigh's term as advocated by Williams. But Williams is the last person who could justly be tarred as a naive "adaptationist." This pejorative was given wide currency by Gould and Lewontin in their overrated "spandrels paper" (1979). It denotes those who assume without evidence that everything an animal is or does must be an adaptation. Unfortunately, their critique of adaptationism has been misunderstood, not least by some philosophers such as the late Jerry Fodor (Daniel Dennett, personal communication), as a critique of the very idea of adaptation itself.

A "spandrel" is a non-adaptive byproduct. The name comes from the gaps between gothic arches which are a necessary but non-functional byproduct of the functionally important arches themselves. Long before the word was introduced into biology, Williams, a leading advocate of adaptation as a proper subject for scientific study, gave an incisive critique of what would later be called spandrels. His vivid example, which regularly grabbed the attention of my Oxford students, was a fox repeatedly running along its own tracks in the snow. Its paws increasingly flattened the snow, which made each successive journey easier and faster. But

x

FOREWORD

it would be wrong to say the fox's paws were adapted to flatten snow. They can't help flattening snow. This particular beneficial effect is a byproduct. Williams summed up the message pithily: adaptation is an "onerous concept."

If I might paraphrase the Anglican marriage service in a way that Williams might not, any attribution of adaptation should not be entered into unadvisedly or lightly; but reverently, discreetly, advisedly, soberly and in the fear of Occam's Razor. You must first assure yourself that you could, if called upon to do so, translate your adaptation theory back into the rigorous terms of Neo-Darwinism. The "adaptation" you postulate must not just be "beneficial" in some vague, panglossian sense. You must clearly set out, and be prepared to defend, a strictly Darwinian pathway to the evolution of the alleged adaptation. The "benefit" must accrue at the proper level in the hierarchy of life which is the unit of Darwinian natural selection. And the proper level, for Williams as for me, is that of the individual genes responsible for the putative adaptation.

"Panglossian" was introduced into biology by JBS Haldane, one of the architects of the Synthesis. His star pupil John Maynard Smith reported that Haldane proposed three "theorems" to satirize errors in scientific thinking.

Aunt Jobiska's Theorem (from Edward Lear): "It's a fact the whole world knows."

The Bellman's Theorem (from Lewis Carroll): "What I tell you three times is true."

Pangloss's Theorem (from Voltaire and applying especially to biology): "All is for the best in the best of all possible worlds."

I adore Sydney Brenner's twinkly-eyed lampoon of panglossism. He imagines a protein, favored in the Cambrian

FOREWORD

because "It might come in handy in the Cretaceous." That's a bit extreme, but my second paragraph above was a list of panglossian errors frequently perpetrated by professors and students alike (including my own undergraduate self). Adaptations cannot be just "good." It is not enough that they convey "benefit." They have to be good for some entity that has been naturally selected precisely because of benefit to itself. And that entity, as Williams powerfully argues, will normally be the gene. I'm fond of a Williams *bon mot* from his later book, *Natural Selection:*

> A gene pool is an imperfect record of a running average of selection pressures over a long period of time in an area often much larger than individual dispersal distances.

But why the gene? And "gene" in what sense? Williams's rationale was so clear and irrefutable, I'm inclined to quote it in full but you only have to turn to page 23, the "Socrates paragraph," which also grabbed my Oxford students by the collar when they read it. Here's the central point.

> With Socrates' death, not only did his phenotype disappear, but also his genotype . . . Socrates' genes may be with us yet, but not his genotype because meiosis and recombination destroy genotypes as surely as death. . . . It is only the meiotically dissociated fragments of the genotype that are transmitted in sexual reproduction, and those fragments are further fragmented by meiosis in the next generation. If there is an ultimately indivisible fragment it is, by definition, 'the gene' that is treated in the abstract discussions of population geneticists.

That last sentence is the answer to my second question, "Gene, in what sense?" I summed up the Williams answer a decade later when I jokingly wrote that *The Selfish Gene*

FOREWORD

might better have been called *The slightly selfish big bit of chromosome and the even more selfish little bit of chromosome.* It could also have been called *The cooperative gene,* and here lies the answer to perhaps the commonest criticism of the "gene's eye view" of natural selection. There is no simple, atomistic, one-to-one mapping between single genes and units of phenotype. Most genes have effects in many parts of the body, and most phenotypic features are influenced by many genes: how then, the critics bleat, can "the gene" be the unit of natural selection? The objection is easily answered and Williams dispatches it with characteristic aplomb:

> No matter how functionally dependent a gene may be, and no matter how complicated its interactions with other genes and environmental factors, it must always be true that a given gene substitution will have an arithmetic mean effect on fitness in any population. (p. 57)

Williams is eloquent on the idea that the other genes in the genome (which in the long run means in the population gene pool) constitute the main environment in which a gene operates—the "background" against which it is naturally selected. The fallacy (a sadly common one) is to assume that a coadapted gene complex is necessarily selected as a unit. Rather, each gene in the complex is selected individually for its compatibility with the other genes in the complex, which are in turn being selected for the very same compatibility.

Return for a moment to Williams' picturesque example of the fox in the snow. I think he'd have accepted the following reservation to his "spandrel" or byproduct lesson. Natural selection actually could favor an adaptive broadening of fox paws for the function of flattening snow. But *only* if the resulting path benefited the fox itself (and its family) alone, rather than foxes in general. It might, for example, be confined to the individual fox's own territory. This brings me to the central core of the book, which is Williams's critique of "Group Selection." This is as needed today as it was in 1966, for group selectionism won't lie down. With its magnetic allure, perhaps politically or even aesthetically motivated, group selectionism keeps coming back for more, in ways that, I can't resist confessing, remind me of Monty Python's Black Knight.

Williams admits that natural selection could theoretically choose among groups. He just doesn't think it's important in practice. His meaning of group selection includes what he later called "clade selection." A hypothetical example might be a tendency for within-species natural selection (which Williams calls "organic selection") to favor larger sized individuals while at the same time whole *species* of smaller individuals are less likely to go extinct ("biotic selection"). Some authors espouse a different form of group selection in which altruistic or cooperative behavior of individuals, or indeed a tendency to live in groups, is thought to be favored because it benefits the group. Williams declines to invoke group selection where the phenomena are more parsimoniously explained by kin selection (Hamilton's seminal papers had just appeared) or reciprocation (Trivers' clever theorizing lay in the future but Williams anticipates the basic idea on page 94). As for living in groups there are, of course, numerous ways in which individuals benefit: huddling for warmth, safety in numbers when predators strike, the "many eyes effect" when spotting opportunities, aerodynamic or hydrodynamic facilitation in flocking birds or schooling fish, "non-zero sum games" in bringing down large prey, etc. Indeed, all these examples are nowadays often handled by Game Theory models in which individuals maximize their own benefit in the context of other individuals maximizing *their* own benefit. Group benefit plays absolutely no part in such models. Incidentally, on page 68 Williams has a prescient

FOREWORD

anticipation of Evolutionary Game Theory, in a slightly different context. Williams revisited and updated his critique of group selection in his 1992 book, *Natural Selection,* but I'll say no more about it here.

In the final chapter of the present book, where he lays out his program for a scientific teleonomy, Williams quotes William Paley's *Natural Theology* on the vertebrate eye. His purpose is to illustrate the self-evident "design" of living creatures, an immensely powerful illusion of design, which pervades some (though not all—that would be adaptationism) biological entities such as the eye. The complex, statistically improbable juxtaposition of mutually-suited functionally cooperating parts—precision-focusing lens, precision-adjusting iris diaphragm, retina with millions of color-coding photocells, optic nerve trunk cable to the brain, such phenomena (and they are legion in all parts of all animals and plants) can only be explained if the principles of chemistry and physics are supplemented by "the one additional postulate of natural selection and its consequence, adaptation." Philosophers and others who don't see the glaring need for natural selection (or divine creation as Paley would have it) must simply be ignorant of the relevant beautiful facts. Have they never seen a David Attenborough film? Or looked down a microscope at a cell? Or contemplated their own hand?

Williams urges us to take seriously the need for a special kind of explanation of adaptation, but to pay cautious attention to the precise mechanism of natural selection and the level in the hierarchy of life where it acts. It is his contention that genic selection occupies the appropriate level. Group selection is a theoretical possibility but it lacks the power to build up Paleyesque complexity: Darwin's "Organs of extreme perfection and complexity," organs which, for Hume, "ravish into admiration all men who have ever contemplated them." We marvel at complex organs that give individuals the power to see, birds to fly, bats to

FOREWORD

echolocate, dogs to smell, cheetahs to sprint. There are no complex organs that give species, or groups, or ecosystems, the power to do anything. Those larger groupings of individuals are just not the kind of entity that has complex "organs" or, indeed, adaptations of any kind. What groups do is a consequence, a byproduct indeed, of what their component individuals do.

George Williams, a tall, imposing, Abe Lincolnesque figure, quiet, kind, thoughtful, modest, made major research contributions to solving outstanding problems in evolutionary biology—really big problems like the evolution of sex, of senescence, of life-history strategies. He was a pioneer of the up-and-coming but still under-valued subject of Darwinian medicine. His *Natural Selection: Domains, Levels and Challenges* was an important successor to this book. But I think *Adaptation and Natural Selection* is his outstanding achievement. When I re-read it before writing this Foreword I expected to find passages that needed critical updating or even deleting. I failed. It can still be recommended to today's students without reserve. Not for its historical interest like some books of the Synthesis, but because this 50-year-old book is still biologically illuminating, wise and—as far as I can judge—correct.

RICHARD DAWKINS, 2018

Preface (1996)

MY FIRST AWARENESS of a motive for *Adaptation and Natural Selection* came during the 1954–55 academic year while on a teaching fellowship at the University of Chicago. The triggering event may have been a lecture by A. E. Emerson, a renowned ecologist and termite specialist. The lecture dealt with what Emerson termed *beneficial death*, an idea that included August Weismann' theory that senescence was evolved to cull the old and impaired from populations so that fitter youthful individuals could take their places. My reaction was that if Emerson's presentation was acceptable biology, I would prefer another calling. Walking home from the lecture with my wife, Doris, I regaled her with my unhappiness about the lecture and proposed the obvious idea that selection among individuals in any population would be biased in favor of the young, as long as the likelihood of living to age x was greater than to age $x + 1$.

A broader dissatisfaction arose from what seemed to be a pervasive inconsistency in the use of the theory of natural selection. Formal presentations of this theory dealt with genetic variation among individuals in a population. Whatever features assisted individuals in their efforts to survive and reproduce would come to characterize the population. No other factor that could produce adaptive change was ever proposed. Yet the adaptations recognized by professional biologists—Emerson, for example—often related to the group as such, and often required

PREFACE (1996)

individual members to jeopardize their own interests for group welfare. Emerson's proposal that individuals adaptively died for the benefit of the population was a typical example.

The simple textbook form of the theory of natural selection was not in fact the only version available. Emerson himself had proposed that selection must operate not only among the individuals of a population, but also among alternative populations (Allee et al., 1948: 664). Others had made the same suggestion, but these were rare exceptions that I had never encountered. I doubt that many of the biologists proposing that there were adaptations for the benefit of the species had them in mind.

Two publications that I read that year were also of great importance to me, and may have deterred me from some alternative career. One was Shaw and Mohler's brief article on sex ratio (1953). The other was David Lack's chapter in Huxley, Hardy, and Ford (1954). Shaw and Mohler's superb work went largely unnoticed until it was discovered by Charnov (1982), who made the Shaw-Mohler equation a focal concept for his broad treatment of sex allocation. Shaw and Mohler led me to Shaw's Ph.D. dissertation, which elaborated the key ideas more fully, and which I cited in the book. Right from its opening paragraph, Lack's "The Evolution of Reproductive Rates" was a sublime encouragement. I had found a biologist who believed, as decisively as I did, that natural selection is a real scientific theory. It logically predicts that there are certain sorts of properties that organisms must have, and others, such as adaptations for the "benefit to the species" (Fisher, 1958: 49–50) that they could not possibly have.

When I was writing the manuscript in the early 1960s, I was convinced of the validity of my position, else why would I have taken the trouble? I was convinced that adaptation was pervasive in biology, essentially defining the

xviii

PREFACE (1996)

subject; that natural selection could explain all examples of adaptation; that adaptations, with few exceptions, were the properties of individual organisms and not groups thereof. I fully expected that the perspective I urged would ultimately be accepted as orthodox; however, I did not at that time expect it to prevail so soon. My impression now is that most biologists who seriously considered the topics I discussed in 1966 had accepted my main ideas by the early 1970s.

I may have been afflicted by a common form of megalomania in assuming that I had the uniquely right perspective and was far ahead of my time. Awareness of the work of Lack and of Shaw and Mohler should have made me doubtful about any such claim. I was really forced to abandon it by the discovery of other publications that anticipated some of what I was writing, especially W. D. Hamilton's 1964 works on inclusive fitness. Hamilton's editorial effect was some last-minute alterations in my Chapters 6 and 7. The psychological effect was a hope that many of my contemporaries might be more receptive to my book than I had been expecting. This hope was abundantly confirmed in the writings of Hamilton, Ghiselin, Maynard Smith, Trivers, and many others in the ten years following the publication of *Adaptation and Natural Selection*.

There are some unfortunate factual and logical flaws in the book. For instance, I did not understand the functioning of a bee sting (pp. 230–31), and it was silly of me to assume that reciprocity demanded advanced cognitive abilities (pp. 95–96). The mistake that I most regretted, soon after the book was published, was my failure to recognize, in the discussion of sexuality (pp. 130–33), the serious theoretical difficulty posed by the cost of meiosis. The chronology of my recollections here is confused, but I do remember noting, as I read the manuscript for John Maynard Smith's "The Origin and

PREFACE (1996)

Maintenance of Sex," that our views were much the same. Maynard Smith's 1971 work had been orphaned by the death of the publication in which it was to appear. I was then editing a multiauthored work for which I thought Maynard Smith's essay eminently suitable. At my request he made it available. According to Ghiselin (1989: 16), I discussed the cost of meiosis in a referee report on a work of his in 1969. It was the central theme of my *Sex and Evolution* (Williams, 1975).

A few years after 1966, I was being given credit for showing that the adaptation concept was not usually applicable at the population or higher levels, and that Wynne-Edwards's thesis that group selection regularly leads to regulation of population density by individual restraints on reproduction was without merit. It also became fashionable to cite my work (sometimes, I suspect, by people who had not read it) as showing that effective selection above the individual level can be ruled out. My recollection, and my current interpretation of the text, especially of Chapter 4, indicate that this is a misreading. I concluded merely that group selection was not strong enough to produce what I termed *biotic adaptation*: any complex mechanism clearly designed to augment the success of a population or a more inclusive group. A biotic adaptation would be characterized by organisms' playing roles that would subordinate their individual interests for some higher value, as in the often proposed benefit to the species.

Even without its producing biotic adaptation, group selection can still have an important role in the evolution of the Earth's biota. The most credible example is the prevalence of sexual reproduction in all the major groups of eukaryote organisms. The phylogenetic distribution of exclusively asexual animals and plants strongly suggests this. There are many asexual species, but a dearth of more inclusive asexual taxa. It seems to happen commonly that

PREFACE (1996)

asexual species are evolved, but then seldom persist long enough to diverge into groups of related species. The early Pleistocene ancestors of today's asexual species were mainly sexuals. The early Pleistocene asexuals have few modern descendants. Phylogenetic survival seems to be biased in favor of those forms that retain sexual reproduction. This view is accepted by most evolutionary biologists today and is a clear use of the group selection concept in explaining an important property of today's biota, its prevalent sexuality.

In evolution it is easier to lose elaborate mechanisms, such as the sexual cycle of gametogenesis and fertilization, than to acquire them. Other likely examples are the loss of eyes and pigmentation in cave animals, of metabolic capabilities in blood parasites, of flight by insular birds and insects, and of planktotrophic larvae by marine invertebrates. These losses would be examples of what Harvey and Partridge (1988) call evolutionary *black holes*. They are paths often taken in evolution, but once taken are largely irreversible.

The fact that most organisms today are not in black holes must be ascribed to a kind of group selection that Stearns (1986) termed *clade selection* and that I have discussed in some detail (Williams, 1992). Evolutionary black holes seem to be traps that often lead to extinction or at least extremely limited opportunity for phylogenetic proliferation. In some cases we can easily understand why. It is no mystery that the forests and grasslands are not teeming with the eyeless descendants of groups that originated in caves. It is less obvious why they are not teeming with asexual animals and plants, although this is a much-investigated problem (Michod and Levins, 1988).

Literature Cited

Allee, W. C., A. E. Emerson, O. Park, T. Park, and K. P.

PREFACE (1996)

Schmidt. 1948. *Principles of animal ecology.* Philadelphia and London: W. B. Saunders.

Charnov, E. L. 1982. *The theory of sex allocation.* Princeton: Princeton University Press.

Fisher, R. A. [1930] 1958. *The genetical theory of natural selection.* Reprint, New York: Dover.

Ghiselin, M. T. 1988. The evolution of sex: A history of competing points of view. Pages 7–23 in *The evolution of sex,* edited by R. E. Michod and B. R. Levin. Sunderland: Sinauer.

Hamilton, W. D. 1964. The genetical evolution of social behavior. Parts 1 and 2. *J. Theoret. Biol.* 7: 1–52.

Harvey, P. H., and L. Partridge. 1988. Murderous mandibles and black holes in hymenopterous wasps. *Nature* 326: 128–29.

Huxley, J. S., A. C. Hardy, and E. B. Ford, editors. 1954. *Evolution as a process.* London: Allen & Unwin.

Maynard Smith, J. 1971. The origin and maintenance of sex. Pages 163–75 in *Group Selection,* edited by G. C. Williams. New York: Aldine-Atherton.

Michod, R. E., and B. R. Levin. 1988. *The evolution of sex.* Sunderland: Sinauer.

Shaw, R. D., and J. D. Mohler. 1953. The selective significance of the sex ratio. *Am. Naturalist* 87: 337–42.

Stearns, S. C. 1986. Natural selection and fitness, adaptation and constraint. Dahlem Konferenzen, Life Sciences Research Report, 36: 23–44.

Williams, G. C. 1975. *Sex and evolution.* Princeton: Princeton University Press.

———. 1992. *Natural selection: Domains, levels, and challenges.* New York and Oxford: Oxford University Press.

Preface

THIS BOOK attempts to clarify certain issues in the study of adaptation and the underlying evolutionary processes. It is directed at advanced students and biologists in general, not at a particular group of specialists. Well-informed readers may find the discussion at an unprofitably elementary level in their own special fields, but they should know where to skip or skim.

The writing dates largely from the summer of 1963, when I had the use of the library of the University of California in Berkeley, California. I am grateful for the aid and cooperation extended by the officials and personnel of this institution. I am also grateful to Miss Jessie E. Miller of Oakland, California, who provided housing for me and my family near the Berkeley campus. Dr. James A. Fowler and Dr. Robert E. Smolker of the State University of New York at Stony Brook made some excellent suggestions on the manuscript.

GEORGE C. WILLIAMS

Stony Brook, Long Island, N.Y.

ADAPTATION AND NATURAL SELECTION
A Critique of Some Current Evolutionary Thought

CHAPTER 1

Introduction

MANY of the contributions to evolutionary thought in the past century can be put in one of two opposed groups. One group emphasizes natural selection as the primary or exclusive creative force. The other minimizes the role of selection in relation to other proposed factors. R. A. Fisher (1930, 1954) showed that many of the proposed alternatives could be discounted with the acceptance of Mendelian genetics and a logical investigation of its relation to selection. Even without Mendelian genetics, Weismann (1904) effectively championed natural selection against some of its rivals of the nineteenth century. His only serious errors are traceable to his ignorance of the Mendelian gene.

The contest was decisively won by natural selection, in my opinion, when by 1932 the classic works of Fisher, Haldane, and Wright had been published. Yet even though this theory may now reign supreme, its realm still supports some opposition, perhaps more than is generally realized. Many recent discussions seem on the surface to conform to the modern Darwinian tradition, but on careful analysis they are found to imply something rather different. I believe that modern opposition, both overt and cryptic, to natural selection, still derives from the same sources that led to the now discredited theories of the nineteenth century. The opposition arises, as Darwin himself observed, not from what reason dictates but from

INTRODUCTION

the limits of what the imagination can accept. It is difficult for many people to imagine that an individual's role in evolution is entirely contained in its contribution to vital statistics. It is difficult to imagine that an acceptable moral order could arise from vital statistics, and difficult to dispense with belief in a moral order in living nature. It is difficult to imagine that the blind play of the genes could produce man. Major difficulties also arise from the current absence of rigorous criteria for deciding whether a given character is adaptive, and, if so, to precisely what is it an adaptation. As I will argue at some length, adaptation is often recognized in purely fortuitous effects, and natural selection is invoked to resolve problems that do not exist. If natural selection is shown to be inadequate for the production of a given adaptation, it is a matter of basic importance to decide whether the adaptation is real.

I hope that this book will help to purge biology of what I regard as unnecessary distractions that impede the progress of evolutionary theory and the development of a disciplined science for analyzing adaptation. It opposes certain of the recently advocated qualifications and additions to the theory of natural selection, such as genetic assimilation, group selection, and cumulative progress in adaptive evolution. It advocates a ground rule that should reduce future distractions and at the same time facilitate the recognition of really justified modifications of the theory. The ground rule—or perhaps *doctrine* would be a better term—is that adaptation is a special and onerous concept that should be used only where it is really necessary. When it must be recognized, it should be attributed to no higher a level of organiza-

INTRODUCTION

tion than is demanded by the evidence. In explaining adaptation, one should assume the adequacy of the simplest form of natural selection, that of alternative alleles in Mendelian populations, unless the evidence clearly shows that this theory does not suffice.

Evolutionary adaptation is a phenomenon of pervasive importance in biology. Its central position is emphasized in the current theory of the origin of life, which proposes that the chemical evolution of the hydrosphere produced at one stage an "organic soup" of great chemical complexity, but lifeless in its earliest stages. Among the complexities was the formation of molecules or molecular concentrations that were autocatalytic in some manner. This is a common chemical property. Even a water molecule can catalyze its own synthesis. Only rarely would a molecule be formed that would produce chance variations among its "offspring" and have such variations passed on to the next "generation," but once such a system arose, natural selection could operate, adaptations would appear, and the Earth would have a biota.

The acceptance of this account of the origin of life implies an acceptance of the key position of the concept of adaptation and at least an abstract criterion whereby life may be defined and recognized. We are dealing with life when we are forced to invoke natural selection to achieve a complete explanation of an observed system. In this sense the principles of chemistry and physics are not enough. At least the one additional postulate of natural selection and its consequence, adaptation, are needed.

This is a very special principle, uniquely biological, and must not be invoked unnecessarily. If asked to

INTRODUCTION

explain the trajectory of a falling apple, given an adequate description of its mechanical properties and its initial position and velocity, we would find the principles of mechanics sufficient for a satisfying explanation. They would be as adequate for the apple as for a rock; the living state of the apple would not make this problem biological. If, however, we were asked how the apple acquired its various properties, and why it has these properties instead of others, we would need the theory of natural selection, at least by implication. Only thus could we explain why the apple has a waterproof wax on the outside, and not elsewhere, or why it contains dormant embryos and not something else. We would find that an impressive list of structural details and processes of the apple can be understood as elements of a design for an efficient role in the propagation of the tree from which it came. We attribute the origin and perfection of this design to a long period of selection for effectiveness in this particular role.

The same story could be told for every normal part or activity of every stage in the life history of every species in the biota of the Earth, past or present. For the same reason that it was once effective in the theological "argument from design," the structure of the vertebrate eye can be used as a dramatic illustration of biological adaptation and the necessity for believing that natural selection for effective vision must have operated throughout the history of the group. In principle, any other organ could be used for illustration although the adaptive design of some parts may not be as immediately convincing as that of the optics of the eye.

This book is based on the assumption that the laws

INTRODUCTION

of physical science plus natural selection can furnish a complete explanation for any biological phenomenon, and that these principles can explain adaptation in general and in the abstract and any particular example of an adaptation. This is a common but not a universal belief among biologists. There are many statements in the recent literature that imply that natural selection can account for some of the superficial forms that adaptation can take, but that adaptation as a general property is something elemental and absolute in living organisms. This is what Russell (1945, p. 3) meant when he said that "directive activity" is an "irreducible characteristic of life." His position was especially clear in his treatment of regeneration, which I will consider on pp. 83-87.

A more recent attack on the adequacy of the simple form of the theory of natural selection is seen in the work of Waddington (1956, 1957, 1959), who conceded that selection is important at every level of adaptive organization, but that it is inadequate by itself and must be supplemented by "genetic assimilation." Another recent attack on the adequacy of natural selection was made by Darlington (1958), who endowed the chromosomes and genes with an evolutionary spontaneity well beyond what is contained in the traditional concept of chance mutation, and with a long-range foresight that allows preparation for the future needs of the population. I will discuss Waddington's and Darlington's work later on.

Even among those who have expressed the opinion that selection is the sole creative force in evolution, there are some inconsistent uses of the concept. With some minor qualifications to be discussed later, it can be said that there is no escape from the conclusion

INTRODUCTION

that natural selection, as portrayed in elementary texts and in most of the technical contributions of population geneticists, can only produce adaptations for the genetic survival of individuals. Many biologists have recognized adaptations of a higher than individual level of organization. A few workers have explicitly dealt with this inconsistency, and have urged that the usual picture of natural selection, based on alternative alleles in populations, is not enough. They postulate that selection at the level of alternative populations must also be an important source of adaptation, and that such selection must be recognized to account for adaptations that work for the benefit of groups instead of individuals. I will argue in Chapters 4 through 8 that the recognition of mechanisms for group benefit is based on misinterpretation, and that the higher levels of selection are impotent and not an appreciable factor in the production and maintenance of adaptation.

DIFFICULTIES in interpretation, especially with respect to the many supposedly group-related adaptations, may result from inappropriate criteria for distinguishing adaptations from fortuitous effects. They are also encouraged by imperfections of terminology. Any biological mechanism produces at least one effect that can properly be called its goal: vision for the eye or reproduction and dispersal for the apple. There may also be other effects, such as the apple's contribution to man's economy. In many published discussions it is not at all clear whether an author regards a particular effect as the specific function of the causal mechanism or merely as an incidental consequence. In some cases it would appear that he has

INTRODUCTION

not appreciated the importance of the distinction. In this book I will adhere to a terminological convention that may help to reduce this difficulty. Whenever I believe that an effect is produced as the function of an adaptation perfected by natural selection to serve that function, I will use terms appropriate to human artifice and conscious design. The designation of something as the *means* or *mechanism* for a certain *goal* or *function* or *purpose* will imply that the machinery involved was fashioned by selection for the goal attributed to it. When I do not believe that such a relationship exists I will avoid such terms and use words appropriate to fortuitous relationships such as *cause* and *effect*. This is a convention in general use already, perhaps unconsciously, and its appropriateness is supported in discussions by Muller (1948), Pittendrigh (1958), Simpson (1962), and others.

Thus I would say that reproduction and dispersal are the goals or functions or purposes of apples and that the apple is a means or mechanism by which such goals are realized by apple trees. By contrast, the apple's contributions to Newtonian inspiration and the economy of Kalamazoo County are merely fortuitous effects and of no biological interest.

It is often easy, in practice, to perceive functional design intuitively, but unfortunately disputes sometimes arise as to whether certain effects are produced by design or merely as by-products of some other function. The formulation of practical definitions and sets of objective criteria will not be easy, but it is a problem of great importance and will have to be faced. An excellent beginning was made by Sommerhoff (1950), but apparently no one has built upon the foundation he provided. In this book I will rely

INTRODUCTION

on informal arguments as to whether a presumed function is served with sufficient precision, economy, efficiency, etc. to rule out pure chance as an adequate explanation.

A frequently helpful but not infallible rule is to recognize adaptation in organic systems that show a clear analogy with human implements. There are convincing analogies between bird wings and airship wings, between bridge suspensions and skeletal suspensions, between the vascularization of a leaf and the water supply of a city. In all such examples, conscious human goals have an analogy in the biological goal of survival, and similar problems are often resolved by similar mechanisms. Such analogies may forcefully occur to a physiologist at the beginning of an investigation of a structure or process and provide a continuing source of fruitful hypotheses. At other times the purpose of a mechanism may not be apparent initially, and the search for the goal becomes a motivation for further study. Adaptation is assumed in such cases, not on the basis of a demonstrable appropriateness of the means to the end but on the indirect evidence of complexity and constancy. Examples are (or were) the rectal glands of sharks, cypress "knees," the lateral lines of fishes, the anting of birds, the vocalization of porpoises.

The lateral line is a good illustration. This organ is a conspicuous morphological feature of the great majority of fishes. It shows a structural constancy within taxa and a high degree of histological complexity. In all these features it is analogous to clearly adaptive and demonstrably important structures. The only missing feature, to those who first concerned themselves with this organ, was a convincing story as to

INTRODUCTION

how it might make an efficient contribution to survival. Eventually painstaking morphological and physiological studies by many workers demonstrated that the lateral line is a sense organ related in basic mechanism to audition (Dijkgraaf, 1952, 1963). The fact that man does not have this sense organ himself, and had not perfected artificial receptors in any way analogous, was a handicap in the attempt to understand the organ. Its constancy and complexity, however, and the consequent conviction that it must be useful in some way, were incentives and guides in the studies that eventually elucidated the workings of an important sensory mechanism.

I have stressed the importance of the use of such concepts as biological means and ends because I want it clearly understood that I think that such a conceptual framework is the essence of the science of biology. Much of this book, however, will constitute an attack on what I consider unwarranted uses of the concept of adaptation. This biological principle should be used only as a last resort. It should not be invoked when less onerous principles, such as those of physics and chemistry or that of unspecific cause and effect, are sufficient for a complete explanation.

For an example that I assume will not be controversial, consider a flying fish that has just left the water to undertake an aerial flight. It is clear that there is a physiological necessity for it to return to the water very soon; it cannot long survive in air. It is, moreover, a matter of common observation that an aerial glide normally terminates with a return to the sea. Is this the result of a mechanism for getting the fish back into water? Certainly not; we need not

INTRODUCTION

invoke the principle of adaptation here. The purely physical principle of gravitation adequately explains why the fish, having gone up, eventually comes down. The real problem is not how it manages to come down, but why it takes it so long to do so. To explain the delay in returning we would be forced to recognize a gliding mechanism of an aerodynamic perfection that must be attributed to natural selection for efficiency in gliding. Here we would be dealing with adaptation.

In this example it would be absurd to recognize an adaptation to achieve the mechanically inevitable. I believe, however, that this is essentially what is done by those who propose that gene mutation is a mechanism for ensuring evolutionary plasticity Whatever the gene may be, it is part of the mundane world and therefore not perfect, in its self-duplication or in any other features. Its occasional alteration is physically inevitable. I will say more about this and related matters on pp. 138-141.

A frequent practice is to recognize adaptation in any recognizable benefit arising from the activities of an organism. I believe that this is an insufficient basis for postulating adaptation and that it has led to some serious errors. A benefit can be the result of chance instead of design. The decision as to the purpose of a mechanism must be based on an examination of the machinery and an argument as to the appropriateness of the means to the end. It cannot be based on value judgments of actual or probable consequences.

This can also be illustrated with an example that will probably not be controversial. Consider a fox on its way to the hen house for the first time after a heavy snowfall. It will probably encounter consider-

INTRODUCTION

able difficulty in forcing its way through the obstructing material. On subsequent trips, however, it may follow the same path and have a much easier time of it, because of the furrow it made the first time. This formation of a path through the snow may result in a considerable saving of time and food energy for the fox, and such savings may be crucial for survival. Should we therefore regard the paws of a fox as a mechanism for constructing paths through snow? Clearly we should not. It is better, because it avoids the onerous biological principles of adaptation and natural selection, to regard the trail-blazing as an incidental effect of the locomotor machinery, no matter how beneficial it may be. An examination of the legs and feet of the fox forces the conclusion that they are designed for running and walking, not for the packing or removal of snow. At any rate, the concept of design for snow removal would not explain anything in the fox's appendages that is not as well or better explained by design for locomotion.

Although the construction of a path through the snow should not be considered a function of the activities that have this effect, the fox does adaptively exploit the effect by seeking the same path on successive trips to the hen house. The sensory mechanisms by which it perceives the most familiar and least obstructed routes and the motivation to follow the path of least effort are clearly adaptations.

Sometimes the important effect of an adaptation from man's point of view may not be its function. Biologically, the brewing of beer is not the function of the glycolytic enzymes of yeast. The production of guano is an important effect, for man's agricultural interests, of the digestive mechanisms of a number of

INTRODUCTION

species of marine birds, but this effect is not their function. Many non-biologists think that it is for their benefit that rattles grow on rattlesnake tails.

I ASSUME that modern biologists would be nearly unanimous in agreeing that the effects mentioned above are not functions. More controversial problems of the same sort will be considered in later chapters. Perhaps one more example can be mentioned here, one that is obviously important but can only be considered rather speculatively. Do we really understand the function of man's cerebral hypertrophy? The importance of this unique character to civilized man is that it enables nearly everyone to learn at least a simple trade; it enables many of us to enjoy good literature and play a fair hand of bridge; and it enables a few to become great scientists, poets, or generals. The human mind has presumably been responsible for analogous benefits for as long as man has had a culturally based society. Despite the arguments that have been advanced (e.g., Dobzhansky and Montague, 1947; Singer, 1962), I cannot readily accept the idea that advanced mental capabilities have ever been directly favored by selection. There is no reason for believing that a genius has ever been likely to leave more children than a man of somewhat below average intelligence. It has been suggested that a tribe that produces an occasional genius for its leadership is more likely to prevail in competition with tribes that lack this intellectual resource. This may well be true in the sense that a group with highly intelligent leaders is likely to gain political supremacy over less gifted groups, but political domination need not result in genetic domination, as is

INTRODUCTION

indicated by the failure of many a ruling class to maintain its numbers. Reasons for questioning the importance of selection between groups in general will be advanced in Chapter 4. Here I will merely note that the close similarity of modern races in their intellectual potentialities would argue against the effectiveness of selection between modern groups as a way of improving man's mental qualities or even of maintaining its present level. The absence of a conspicuous decline in human mentality within historical time must mean that selection has somehow continued to promote the survival of the intelligent.

I suggest that advanced mental qualities might possibly be produced as an incidental effect of selection for the ability to understand and remember simple verbal instructions early in life. We might imagine that Hans and Fritz Faustkeil are told on Monday, "Don't go near the water," and that both go wading and are spanked for it. On Tuesday they are told, "Don't play near the fire," and again they disobey and are spanked. On Wednesday they are told, "Don't tease the saber-tooth." This time Hans understands the message, and he bears firmly in mind the consequences of disobedience. He prudently avoids the saber-tooth and escapes the spanking. Poor Fritz escapes the spanking too, but for a very different reason. Even today accidental death is an important cause of mortality in early life, and parents who consistently spare the rod in other matters may be moved to violence when a child plays with electric wires or chases a ball into the street. Many of the accidental deaths of small children would probably have been avoided if the victims had understood and remembered verbal instructions and had been capable of

INTRODUCTION

effectively substituting verbal symbols for real experience. This might well have been true also under primitive conditions. The resulting selection for acquiring verbal facility as early as possible might have produced, as an allometric effect on cerebral development, populations in which an occasional Leonardo might arise. This interpretation is supported by the apparent diversity of adult human mentality and by the close similarity of the sexes in intellectual endowment. A character selected for a specific adult-male function, such as the political leadership of a primitive society, would reach a high development in adult males only, and would be well standardized in this group. These arguments do not really support my interpretation very strongly, and if anyone has a better theory, I hope he will let it be known, because the problem is surely an important one. Is it not reasonable to anticipate that our understanding of the human mind would be aided greatly by knowing the purpose for which it was designed?

BENEFITS to groups can arise as statistical summations of the effects of individual adaptations. When a deer successfully escapes from a bear by running away, we can attribute its success to a long ancestral period of selection for fleetness. Its fleetness is responsible for its having a *low probability* of death from bear attack. The same factor repeated again and again in the herd means not only that it is a herd of fleet deer, but also that it is a fleet herd. The group therefore has a *low rate* of mortality from bear attack. When every individual in the herd flees from a bear, the result is effective protection of the herd.

INTRODUCTION

As a very general rule, with some important exceptions, the fitness of a group will be high as a result of this sort of summation of the adaptations of its members. On the other hand, such simple summations obviously cannot produce collective fitness as high as could be achieved by an adaptive organization of the group itself. We might imagine that mortality rates from predation by bears on a herd of deer would be still lower if each individual, instead of merely running for its life when it saw a bear, would play a special role in an organized program of bear avoidance. There might be individuals with especially well-developed senses that could serve as sentinels. Especially fleet individuals could lure bears away from the rest, and so on. Such individual specialization in a collective function would justify recognizing the herd as an adaptively organized entity. Unlike individual fleetness, such group-related adaptation would require something more than the natural selection of alternative alleles as an explanation.

It may also happen that the incidental effects of individual activities, of no functional significance in themselves, can have important statistical consequences, sometimes harmful, sometimes beneficial. The depletion of browse is a harmful effect of the feeding activities of each member of a dense population of deer. If browse depletion were beneficial, I suspect that someone, sooner or later, would have spoken of the feeding behavior of deer as a mechanism for depleting browse. A statement of this sort should not be based merely on the evidence that the statistical effect of eating is beneficial; it should be based on an examination of the causal mechanisms to determine whether they cannot be adequately ex-

INTRODUCTION

plained as individual adaptations for individual nourishment.

The feeding activities of earthworms would be a better example, because here the incidental statistical effects *are* beneficial, from the standpoint of the population and even of the ecological community as a whole. As the earthworm feeds, it improves the physical and chemical properties of the soil through which it moves. The contribution from each individual is negligible, but the collective contribution, cumulative over decades and centuries, gradually improves the soil as a medium for worm burrows and for the plant growth on which the earthworm's feeding ultimately depends. Should we therefore call the causal activities of the earthworm a soil-improvement mechanism? Apparently Allee (1940) believed that some such designation is warranted by the fact that soil improvement is indeed a result of the earthworm's activities. However, if we were to examine the digestive system and feeding behavior of an earthworm, I assume that we would find it adequately explained on the assumption of design for individual nutrition. The additional assumption of design for soil improvement would explain nothing that is not also explainable as a nutritional adaptation. It would be a violation of parsimony to assume both explanations when one suffices. Only if one denied that some benefits can arise by chance instead of by design, would there be a reason for postulating an adaptation behind every benefit.

On the other hand, suppose we did find some features of the feeding activities of earthworms that were inexplicable as trophic adaptations but were exactly what we should expect of a system designed

INTRODUCTION

for soil improvement. We would then be forced to recognize the system as a soil-modification mechanism, a conclusion that implies a quite different level of adaptive organization from that implied by the nutritional function. As a digestive system, the gut of a worm plays a role in the adaptive organization of that worm and nothing else, but as a soil-modification system it would play a role in the adaptive organization of the whole community. This, as I will argue at length in later chapters, is a reason for rejecting soil-improvement as a purpose of the worm's activities if it is possible to do so. Various levels of adaptive organization, from the subcellular to the biospheric, might conceivably be recognized, but the principle of parsimony demands that we recognize adaptation at the level necessitated by the facts and no higher.

It is my position that adaptation need almost never be recognized at any level above that of a pair of parents and associated offspring. As I hope to show in the later chapters, this conclusion seldom has to rest on appeals to parsimony alone, but is usually supported by specific evidence.

The most important function of this book is to echo a plea made many years ago by E. S. Russell (1945) that biologists must develop an effective set of principles for dealing with the general phenomenon of biological adaptation. This matter is considered mainly in the final chapter.

CHAPTER 2

Natural Selection, Adaptation, and Progress

ONE of the strengths of scientific inquiry is that it can progress with any mixture of empiricism, intuition, and formal theory that suits the convenience of the investigator. Many sciences develop for a time as exercises in description and empirical generalization. Only later do they acquire reasoned connections within themselves and with other branches of knowledge. Many things were scientifically known of human anatomy and the motions of the planets before they were scientifically explained.

The study of adaptation seems to show the opposite mode of development. It has already had its Newtonian synthesis, but its Galileo and Kepler have not yet appeared. The "Newtonian synthesis" is the genetical theory of natural selection, a logical unification of Mendelism and Darwinism that was accomplished by Fisher, Haldane, and Wright more than thirty years ago. For all its formal elegance, however, this theory has provided very limited guidance in the work of biologists. Ordinarily it does little more than to give a vague aura of validity to conclusions on adaptive evolution and to enable a biologist to refer to goal-directed activities without descending into teleology. The inherent strength of the theory is restricted by the paucity of generalizations, analogous to Kepler's laws, that can serve on the one hand as summaries of large masses of observations and, on the other hand, as logical deductions from the theory.

NATURAL SELECTION, ADAPTATION & PROGRESS

The deficiency of course is not absolute. The kind of generalization I have in mind is well illustrated by Lack's conclusion on the selection of fecundity in animals that feed their young (discussed on pp. 161-162) and Fisher's conclusion on population sex ratios (see pp. 146-156). With perhaps another hundred such insights we could have a unified science of adaptation.

The current lack of such unification has some unfortunate consequences. One is that a biologist can make any evolutionary speculation seem scientifically acceptable merely by adorning his arguments with the forms and symbols of the theory of natural selection. Thus we have biologists recognizing, in the name of natural selection, mutation, isolation, etc., adaptations designed to meet the demands of geologically future events. This fallacy commonly occurs in the guise of provisions for "evolutionary plasticity." Other biologists speak of natural selection as ensuring that an individual or a population will have all the adaptations that are *necessary* for its survival and imply that adaptations are never expected to be more or less than *adequate* to ensure survival. Such powers might appropriately be attributed to a prescient Providence, but certainly not to natural selection, as this process is commonly described.

Another tendency that survives, despite its lack of a theoretical justification, is a belief in a deterministic succession of evolutionary stages. Simpson's book of 1944 can be taken to symbolize the end of orthogenetic interpretations of paleontological data, but long-term evolutionary determinism is still detectable in some discussions of *progress* in evolution. Huxley (1953, 1954), for example, argued that evolutionary progress was inevitable and proceeded by a series of advances to new levels until all possible levels but one had been achieved: "... by the Pliocene only one path of progress remained open—that which led to man" (1954, p. 11). Huxley admits that the details of the process of progressing to higher levels would have been unpredictable at any one point in geological time, but says, "On the other hand, once we can look back on the facts we realize that it could have happened in no other way" (1953, p. 128). The force that drives and guides evolutionary progress is said to be natural selection. This argument is an excellent example of how one can abide by the outward forms of the theory but violate its spirit.

I doubt that many biologists subscribe to the view of evolution as a deterministic progression towards man, but there is widespread belief in some form of aesthetically acceptable progress as an inevitable outcome of organic evolution. In this chapter I will discuss some of the limitations of the process of natural selection and their bearing on some common suppositions, such as the inevitability of progress. The stress on limitations does not indicate any doubt on my part as to the importance of natural selection. Within its limited range of activity, it has a potency that may still be generally underestimated by the majority of biologists. There is a very illuminating discussion by Muller (1948) on this point.

THE ESSENCE of the genetical theory of natural selection is a statistical bias in the relative rates of survival of alternatives (genes, individuals, etc.). The effectiveness of such bias in producing adaptation is contingent on the maintenance of certain quantitative relationships among the operative factors. One necessary condition is that the selected entity must have a high degree of permanence and a low rate of endogenous change, relative to the degree of bias (differences in selection coefficients). Permanence implies reproduction with a potential geometric increase.

Acceptance of this theory necessitates the immediate rejection of the importance of certain kinds of selection. The natural selection of phenotypes cannot in itself produce cumulative change, because phenotypes are extremely temporary manifestations. They are the result of an interaction between genotype and environment that produces what we recognize as an individual. Such an individual consists of genotypic information and information recorded since conception. Socrates consisted of the genes his parents gave him, the experiences they and his environment later provided, and a growth and development mediated by numerous meals. For all I know, he may have been very successful in the evolutionary sense of leaving numerous offspring. His phenotype, nevertheless, was utterly destroyed by the hemlock and has never since been duplicated. If the hemlock had not killed him, something else soon would have. So however natural selection may have been acting on Greek phenotypes in the fourth century B.C., it did not of itself produce any cumulative effect.

The same argument also holds for genotypes. With Socrates' death, not only did his phenotype disappear, but also his genotype. Only in species that can maintain unlimited clonal reproduction is it theoretically possible for the selection of genotypes to be an important evolutionary factor. This possibility is not likely to be realized very often, because only rarely would individual clones persist for the immensities of time that are important in evolution. The loss of Socrates' genotype is not assuaged by any consideration of how prolifically he may have reproduced. Socrates' genes may be with us yet, but not his genotype, because meiosis and recombination destroy genotypes as surely as death.

It is only the meiotically dissociated fragments of the genotype that are transmitted in sexual reproduction, and these fragments are further fragmented by meiosis in the next generation. If there is an ultimate indivisible fragment it is, by definition, "the gene" that is treated in the abstract discussions of population genetics. Various kinds of suppression of recombination may cause a major chromosomal segment or even a whole chromosome to be transmitted entire for many generations in certain lines of descent. In such cases the segment or chromosome behaves in a way that approximates the population genetics of a single gene. In this book I use the term *gene* to mean "that which segregates and recombines with appreciable frequency." Such genes are potentially immortal, in the sense of there being no physiological limit to their survival, because of their potentially reproducing fast enough to compensate for their destruction by external agents. They also have a high degree of qualitative stability. Estimates of mutation rates range from about 10^{-4} to 10^{-10} per generation. The rates of selection of alternative alleles can be much higher. Selection among the progeny of individuals heterozygous for recessive lethals would eliminate half the lethal genes in one generation. Aside from lethal and markedly deleterious genes in experimental populations, there is abundant evidence (e.g.,

NATURAL SELECTION, ADAPTATION & PROGRESS

Fisher and Ford, 1947; Ford, 1956; Clarke, Dickson, and Sheppard, 1963) for selection coefficients in nature that exceed mutation rates by one to many multiples of ten. There can be no doubt that the selective accumulation of genes can be effective. In evolutionary theory, a gene could be defined as any hereditary information for which there is a favorable or unfavorable selection bias equal to several or many times its rate of endogenous change. The prevalence of such stable entities in the heredity of populations is a measure of the importance of natural selection.

Natural selection would produce or maintain adaptation as a matter of definition. Whatever gene is favorably selected is better adapted than its unfavored alternatives. This is the reliable outcome of such selection, the prevalence of well-adapted genes. The selection of such genes of course is mediated by the phenotype, and to be favorably selected, a gene must augment phenotypic reproductive success as the arithmetic mean effect of its activity in the population in which it is selected. Chapter 3 will deal more fully with the connections between a gene and its phenotype and external environment. Chapter 4 will consider more inclusive systems than the gene as objects of natural selection.

A thorough grasp of the concept of a gene's mean phenotypic effect on fitness is essential to an understanding of natural selection. If individuals bearing gene A replace themselves by reproduction to a greater extent than those with gene A', and if the population is so large that we can rule out chance as the explanation, the individuals with A would be, as a group, more fit than those with A'. The difference in their total fitness would be measured by the extent of replacement of one by the other. By definition of mean, the mean effect on individual fitness of A would be favorable and of A' unfavorable. This maximization of mean individual fitness is the most reliable phenotypic effect of selection at the genic level, but even here there are complications and exceptions. For example, a gene might be favorably selected, not because its phenotypic expression favors an individual's reproduction, but because it favors the reproduction of close relatives of that individual. This complication is considered on pp. 195-197. Wright (1949) and Hamilton (1964A) have provided generally applicable theoretical discussions of the relationship of selection to individual fitness.

Natural selection commonly produces fitness in the vernacular sense. We ordinarily expect it to favor mechanisms leading to an increase in health and comfort and a decrease in danger to life and limb, but the theoretically important kind of fitness is that which promotes ultimate reproductive survival. Reproduction always requires some sacrifice of resources and some jeopardy of physiological well-being, and such sacrifices may be favorably selected, even though they may reduce fitness in the vernacular sense of the term.

We ordinarily expect selection to produce only "favorable" characters, but here again there are exceptions. In the effects of a gene there may be influences on more than one character. A given gene substitution may have one favorable effect and another unfavorable one in the same individual, often, but not necessarily, in different parts of the life cycle. The same gene may produce mainly favorable effects in one individual but mainly unfavorable effects in another, because of differences in environment or genetic background. If the mean effect is favorable, the gene will increase in frequency, and so will all its effects, both positive and negative. There are many relevant examples. An embryonic lethality is a character that has been produced in certain mouse populations by natural selection. The gene that causes this condition is favorably selected, up to an appreciable frequency, because of a favorable effect, "meiotic drive" in the male gamete stage (Lewontin and Dunn, 1960). Senescence, certain kinds of "normal" sterility (see Chapters 7 and 8), and various hereditary diseases are other examples of unfavorable characters that owe their prevalence to natural selection. In all such examples, the favorable selection of the genetic basis for such deleterious effects must be ascribed to other effects of the same genes. Favorable selection of a gene is inevitable if it has a favorable mean effect compared to the available alternatives of the moment.

Another frequent outcome of natural selection is the promotion of the long-term survival of the population. One example, the maintenance of fleetness in deer, was cited in the first chapter, and many similar examples could be given. Here again, however, there are exceptions. The constant maximization of mean fitness in some populations might bring about an increasing ecological specialization, and this might mean reduced numbers, restricted range, and vulnerability to changed conditions. Haldane (1932) mentioned flower specialization for very efficient pollination by a taxonomically small group of insects as an example of such vulnerability to extinction caused by natural selection. Haldane also mentioned the production of elaborate weapons or of conspicuous ornamentation and display, which might be favored in competition for mates, as factors that decrease population fitness by the wasteful use of resources and the damage and vulnerability to predators caused by sexual conflict. Probably most evolutionary increases in body size cause a decrease in numbers, and this might contribute to extinction. An excellent example of decrease in numbers brought about by natural selection is the evolution of the slave-making instinct in certain groups of ants (Emerson, 1960).

IN DISCUSSIONS of the role of adaptation in the survival of populations one often finds statements to the effect that selection caused certain developments because they were necessary. It is often difficult to distinguish semantic and conceptual difficulties, but I believe that there are common conceptual fallacies such as might be illustrated by this statement:

> The white coat of the polar bear is *necessary* for the stalking of game in the snowy regions in which it lives. The whiteness was favored by selection because darker individuals were unable to survive.

I would correct this argument by substituting *advantageous* for *necessary* in the first sentence, and by adding the words *as well* to the end of the second. Ecological or physiological necessity is not an evolutionary factor, and the development of an adaptation is no evidence that it was necessary to the survival of the species. We might indulge at this point in the fanciful act of rendering all present polar bears and their descendants a bright pink. We can now be sure that the species will not henceforth survive *as well*. Its numbers will suddenly decline and its geographic and ecological range rapidly contract, but we cannot be sure that this decrease will proceed all the way to extinction. Each polar bear, after meeting unaccustomed frustration in its hunting, will adapt by hunting for longer periods of time. Some may learn that they can hunt more successfully at night than by day. These and other adjustments might enable the species to continue in those regions where pinkness is, for one reason or another, less of a handicap than in others. Needless to say, there are many obviously necessary adaptations. If, instead of depriving the bear of its whiteness we deprived it of its lungs, it would immediately become extinct. Such examples, however, do not invalidate the conclusion that the mere presence of an adaptation is no argument for its necessity, either for the individual or the population. It is evidence only that during the evolutionary development of the adaptation the genes that augmented its development survived *at a greater rate* than those that did not. Usually, but not always, the presence of an adaptation causes the species to be more numerous and widespread than it would be without it. Nicholson (1956, 1960) has discussed this relationship of natural selection to population density and has concluded that improved adaptation would often have but slight effect on numbers, because even slight increases might greatly intensify the density-governed reactions that normally check population growth. Nicholson is the leading champion of the belief that population densities in nature represent stable equilibria.

The converse argument also holds. The fact that

a certain adaptation is necessary to the survival of a species has no bearing on its likelihood of evolving. We can say of every group of organisms that is now extinct that whatever adaptations were necessary for its survival were not, in fact, evolved. This does not demonstrate that there were no tendencies in the necessary direction; it merely means that these tendencies, if there, were not adequate. However, there is no necessity for believing that they were there. The imminence of extinction does not evoke emergency measures on the part of a population. I can imagine that a sonar system would be an advantage in the nocturnal navigation of owls, just as it is for bats. I presume also that many populations of owls have become extinct and that some of these might have survived if provided with even a slight additional advantage, such as a rudimentary sonar system. Would we be more likely to see the beginnings of such a system, or of any other adaptive mechanism, in a small population declining towards extinction than in a large and expanding one? I doubt that any ornithologist would be willing to devote much time looking into such a possibility. I assume that the failure of owls to evolve sonar results from their lack of some necessary preadaptations in all their populations, regardless of size. The lack of sonar is no evidence that it is not necessary for continued existence. Perhaps a post-Recent adaptive radiation of bats will make it necessary for all owls to have an effective sonar system. If so, they will simply join the pterodactyls and hosts of other organisms that lacked necessary adaptations.

The possibility that populations can take special steps in response to the threat of imminent extinction is often implied in elementary biology texts in discussions of adaptive radiation or of the continued survival of ancient types. Certain species, we are told, were able to avoid extinction by seeking marginal habitats, thereby escaping competition from more progressive forms. The avoidance of extinction might well be a result of specialization for niches in which competition is minimal, but it cannot, historically, have been a cause of evolutionary change. Only in an endlessly recurring cycle, as is shown by the succession of generations in a population, can one class of events be both the cause and the effect of another. A mouse can retreat to a hole to avoid being killed by a cat, but a population cannot retreat to a marginal habitat to avoid being killed off by competition. Such a development can only be a secondary effect of the differences in the genetic survival of individuals in the evolving population.

This hypothetical discussion has at least one experimental parallel. Park and Lloyd (1955) showed that experimental flour-beetle populations failed to show a genetic change calculated to avoid the extinction that usually followed the introduction of a competing species. The authors, however, were reluctant to generalize and to conclude that ecological necessity never influences evolutionary change.

There is no way in which a factor of necessity-for-survival could influence natural selection, as this process is usually formulated. Selection has nothing to do with what is necessary or unnecessary, or what is adequate or inadequate, for continued survival. It deals only with an immediate better-vs.-worse within a system of alternative, and therefore competing, entities. It will act to maximize the mean reproductive performance regardless of the effect on long-term population survival. It is not a mechanism that can anticipate possible extinction and take steps to avoid it.

I have indicated above that natural selection works only among competing entities, but it is not necessary for the individuals of a species to be engaged in ecological competition for some limited resource. This requisite is often assumed, beginning with Darwin and continuing with many modern biologists. A little reflection, however, will indicate that natural selection may be most intense when competitive interactions are low. Consider a typical growth curve in an experimental population. Suppose that the genetic variation present in such a population is mainly in the ability to convert food materials, present in excess of maintenance requirements, into offspring. There will then be intense selection during the initial stages of population growth when food is present in excess. Then when competition for food becomes intense, the genetic variation will lose its expression and selection will cease. This is true of all situations in which variation in fitness has greater expressivity under conditions of lower population density. Fuller discussions of this point are provided by Haldane (1931), Birch (1957), Mather (1961), and Milne (1961).

Thus natural selection may be operative in the complete absence of competition in the usual sense. In most animal populations there is no competition for oxygen. The fact that dog A gets enough oxygen in no way influences dog B's efforts to get his share. Only very indirectly, by contributing to other functions, such as food-getting, would respiration relate to ecological competition. The elaborateness and precision of the canine respiratory system, however, leaves no room for doubt that there has been a relentless selection for respiratory efficiency. On the other hand, no organism can escape what I would call reproductive competition. Suppose dog A successfully sires three puppies this year. How can this amount of success be evaluated? Only in comparison with dog B. If neither B nor any other dog in the pack produced more than two, we would conclude that A succeeded very well in reproduction, but if the pack mean were four, that A fared badly. The situation is as clear with competing alleles as with competing individuals. Suppose the gene *a* is present in a thousand individuals in a population in both the present generation and in the preceding one, and the rest of the population has only A. Is *a* being favorably selected? The answer, of course, cannot be given until "the rest" is translated into definite numerical values. If the population is increasing, gene *a* is being unfavorably selected. If it is decreasing, the selection of *a* is favorable.

In its ultimate essence the theory of natural selection deals with a cybernetic abstraction, the gene, and a statistical abstraction, mean phenotypic fitness. Such a theory can be immensely interesting to those who have a liking and a facility for cybernetics and statistics. Fruitful applications of this theory will also require a detailed knowledge of biology. The theory certainly has little appeal to those who are not in the regular habit of using mathematical abstractions in their thinking about organisms. I doubt that the concept of mean reproductive success as applied to honey bees would have much appeal to such people. The bee population is composed of very many quite

NATURAL SELECTION, ADAPTATION & PROGRESS

sterile individuals and a few extraordinarily fertile ones. A bee of mean fertility would be quite atypical. Yet I believe that the sterility of the workers is entirely attributable to the unrelenting efforts of Darwin's demon to maximize a mere abstraction, the mean. As indicated on pp. 195-201, I see little hope of explaining the societies of the social insects in any other way.

DOES this theory, which I believe to be accurately summarized above, necessitate a belief in progress? Many biologists have stated that it does, and many more have tacitly assumed this position. I would maintain, however, that there is nothing in the basic structure of the theory of natural selection that would suggest the idea of any kind of cumulative progress. An organism can certainly improve the precision of its adaptation to current circumstances. It must often happen that a possible allele at a certain locus would be better adapted than any that actually prevail at that locus. The necessary mutation may never have occurred, or may always have been lost by drift when it did occur. Sooner or later, however, such a gene may drift to a sufficiently high frequency to be effectively selected. When this happens it will soon become the normal allele or at least one of a series of normal alleles at its locus. This gene substitution would slightly improve the precision of one or more adaptations, but as perfection is approached the opportunity for further improvement would correspondingly diminish. This is certainly not a process for which the term "progress" would be at all appropriate.

Gene substitution may also take place because of a

changing environment. A gene may at one time be in low frequency because it is unfavorably selected in the environment in which the population finds itself. After a change in environment the gene may be favorably selected and replace its alternative alleles in whole or in part. The phenotypic effect of such a substitution would be to improve one adaptation, one that has increased in importance, at the expense of another that has become less important. This again is not something that would suggest the term "progress," but such selective gene substitution is the only expected outcome of natural selection. I suspect that no one would ever have deduced progress from the theory itself. The concept of progress must have arisen from an anthropocentric consideration of the data bearing on the history of life.

Progress has meant different things to different people. The views are not all mutually exclusive, but for convenience I will consider progress in five separate categories: as accumulation of genetic information; as increasing morphological complexity; as increasing physiological division of labor; as any evolutionary tendency in some arbitrarily designated direction; as increased effectiveness of adaptation.

The most notable contribution on progress as accumulation of information is Kimura's (1961). His discussion is an important contribution to evolutionary theory, but I believe that its value is compromised by his acceptance of some naïve preconceptions. He assumes, without any explicit evidence, that modern zygotes contain more information than Cambrian zygotes. He clearly implies that a human zygote has a higher information content than any other zygote, or at least than most other zygotes. There is no direct

NATURAL SELECTION, ADAPTATION & PROGRESS

evidence on which this position might be contested, but there is also little evidence in its favor. Man is superlative in many ways; he is by far the most intelligent organism, and he has very recently achieved an ecological dominance of unprecedented scope. It does not follow from this that he is superlative in all important respects, such as the measure of negentropy in his genetic code. Genetic codes have changed enormously since Cambrian time, and natural selection has guided these changes, but it need not have increased the total information content.

Kimura's analysis shows that initially nonsensical DNA can be rapidly organized by natural selection so as to constitute instructions for adaptation. He estimates that 10^8 bits of information could be accumulated in a line of descent in the time elapsed since the Cambrian period by natural selection acting in opposition to randomizing forces. He notes that the amount of DNA in a cell may be ten to a hundred times the amount necessary to convey this information, and he interprets this and other evidence to indicate that the DNA message is very redundant. Evidently Kimura believes that a large proportion of the information in the genotype of a higher organism has accumulated since the Cambrian period, and been "written" in then unspecified DNA.

Kimura's discussion is of value in indicating what natural selection can accomplish in half a billion generations. My inclination would be to accept his account of the accumulation of information but to apply it not to the Cambrian but to the period immediately after the first appearance of cellular organisms under the control of a DNA coding system. At that time genetic information would be expected to accumulate, but it is unreasonable to assume that it would accumulate indefinitely. A certain amount of information is added by selection every generation. At the same time, a certain amount is subtracted by randomizing processes. The more information is already stored, the more would mutation and other random forces reduce it in a given time interval. It is reasonable to suppose that there would be a maximum level of information content that could be maintained by selection in opposition to randomizing forces. It also seems inevitable that much of the gene substitution that is actually in progress at any one time is of a readily reversible sort. Kimura made allowances for the neutralization of some of the effect of selection by mutation pressure, and showed that some effect would remain. He assumed that all of this remainder would be available for progress in the accumulation of information, and made no allowances for the reversal of selection. If selection were to change a gene frequency from 0.2 to 0.8 in a century and then in the following century change it back to 0.2, there would have been evolution, but no net accumulation of information.

Sheppard (1954) reviewed evidence that even strong selection pressures in natural insect populations may change sign as a result of slight environmental changes. Selection of isoalleles, with very slight differences in selection coefficients, would be expected to change sign more readily than the major gene differences studied by Sheppard. Reversing selection at a gene locus would not imply a reversal of the course of evolution for the population. Reversals at different loci would be at least partly independent, and the population would have a unique set of gene frequencies at any one moment. In environments that were changing, however, much of natural selection would be directed at undoing what it had recently established. In stable environments selection would usually produce a decrease of heterozygosity (Lewontin, 1958A), and this would also reduce the total genetic information. If one takes these factors into account, Kimura's calculations can be reinterpreted to mean that no genetic information is accumulating in modern organisms.

This conclusion is supported by other considerations. The accumulation of information is a function of selection pressures and numbers of generations. If man's ancestors, as Kimura suggested, have had an average generation length of one year since Cambrian time, and if much of his ancestry before that time was at the protistan stage, there might have been a thousand times as many generations before the Cambrian as since. If so, Kimura's calculations would indicate that 10^{11} bits of information might have accumulated before the Cambrian. There is not enough DNA in a human zygote to carry nearly this amount. I would suspect that long before the Cambrian most phylogenetic lineages had established an optimum amount of DNA and had fully optimized its burden of information. This conclusion is supported by Blum's (1963) reasoning on the rate of approach to perfection in adaptation in a constant environment. His thinking was quite different from Kimura's and amounts to an estimate of the proportion of untried alleles at each locus after a given period of time. The conclusion depends on estimates of mutation rate per generation, but with an average mutation rate of 10^{-6} there would be very little possible progress remaining after 10^7 generations, or perhaps a thousand years in a protist population. Only a short time should be required to establish an equilibrium between the increase of information by natural selection and its destruction by random processes. These processes should be understood to include not only mutation, but also any environmental changes that result in reversals of selection pressures.

It might seem reasonable to expect that the amount of information could be simply measured as the amount of DNA in the zygote. This would be true if we could be sure that the amount of redundancy per unit of zygotic information is constant, but I know of no basis for even indulging in guesswork on this point. Our information on DNA content is meagre and little advanced beyond the beginnings made by Mirsky and Ris (1951) and Vendrely (1955). Such simple protists as the bacteria and sporozoa have small amounts of DNA, as might be expected from the minute size of their cells. The invertebrates are highly variable but with some association between the amount of DNA and position on the phylogenetic scale. This association is largely a function of an extremely high value for the squid, greater than for any mammal. Among the vertebrates the only suggested trend is for a reduction of DNA in the lungfish-to-bird sequence. Mammals have more than birds, but not as much as amphibians. Such inconsistencies between DNA content and the presumed level of evolutionary advance is customarily explained by assuming a high level of genetic redundancy in the lower organisms (Mirsky and Ris, 1951; Waddington, 1962, pp. 59-60). Our knowledge of phylogenetic variation in cellular DNA content is based for the most part on the grossly inadequate evidence of single values for single representatives of phyla and classes. We have little notion of what proportion of the human zygotic information is concerned with structure at the anatomical and histological levels, and what proportion relates to cellular and biochemical mechanisms, in other words, the number of constraints imposed by selection for gross structural adaptations in comparison with those imposed by selection for cellular and biochemical characters. If we knew that three-fourths of the information in the human germ plasm were devoted to morphogenetic instructions, we might conclude that man must have about four times as much zygotic information as an amoeba. But suppose only one-tenth were concerned with morphology and the rest with biochemistry. Man would then have only a little more than an amoeba, and an alga, with its elaborate synthetic enzyme systems, might have more than either.

This is a problem that can be approached only in the most speculative manner, but some of the considerations that Kimura raised have bearing on how much information is present in the zygote. He reasoned that about 10^7 bits of information would be necessary to specify human anatomy, and that a maximum of 10^{10} bits could be carried by the DNA present in the zygote. If we assume that the DNA message is so redundant as to be utilized at only one-tenth of its maximum capacity, we could still conclude from these estimates that a hundred times as much of the information in the germ plasm is concerned with basic cellular and biochemical mechanisms as is concerned with morphogenesis. I suggest this, not as a conclusion to be seriously considered, but as one end of a spectrum of possibilities. The spectrum is so wide that it would seem premature to base any conclusions on an estimate of the difference in genetic information content of a mammal and a protist.

Presumably the amount of DNA would always be regulated at some optimum value by selection. The DNA present in the body would very seldom be a significant mechanical or nutritional burden. If increasing DNA content would permit the carrying of more information and thereby allow greater precision or versatility of adaptation, I presume that such an increase would take place. Economy and efficiency are universal characteristics of biological mechanisms, and the DNA coding system should certainly not be an exception. Its manifest purpose is the carrying of information, and it is reasonable to assume that it is in relation to this function that its quantity is optimized. The optimum would be determined by the amount of information that can be maintained by selection in the face of randomizing processes. The more micrograms of DNA are present, the less will selection be able to control the information content per microgram. Decreased control means increased noise and consequent reduction in the adaptive precision of the phenotype. Quantity of information and precision of information are somewhat opposed requirements of the genetic message. The amount of DNA in an organism would presumably reflect the optimum compromise between these opposing values. It would follow from this that genetic information is limited in amount and must be utilized as economically as possible. A number of biological phenomena, as will be discussed in various contexts in later chapters, suggest that some such principle of economy of information is an important evolutionary factor. The view suggested here is that all organisms of above a certain low level of organization—perhaps that of the simpler invertebrates—and beyond a certain geological period—perhaps the Cambrian—may have much the same amounts of information in their nuclei. All such organisms have quantities of DNA capable of carrying an enormous amount of information, and all have ancestries of at least 10^9 years and an astronomical number of generations that were subject to the same information-generating force. We can interpret evolution since the Cambrian as a history of substitutions and qualitative changes in the germ plasm, not an increase in its total content. Evolution from protist to man might have been largely a matter of the substitution of morphogenetic instructions for a small proportion of the biochemical and cytological instructions in the protist DNA.

KIMURA's conclusions on cybernetic progress arose from his acceptance of the second category of evolutionary progress, that of increasing morphological complexity. It is often stated or implied that animals of the Recent epoch are morphologically more complex than those of the Paleozoic era, but I am not aware of any objective and unbiased documentation of this point. Is man really more complex structurally than his piscine progenitor of Devonian time? We can certainly describe a more complex series of evolutionary changes in, for example, the human skull than in the Devonian fish skull, but this is at least

NATURAL SELECTION, ADAPTATION & PROGRESS

partly attributable to our ignorance of pre-Devonian chordates. The Devonian-to-Recent lineage of man is mainly a history of changing arrangements and losses of parts, in the skull and elsewhere. Real additions are not a conspicuous part of the story. Mechanically the human skull is exceedingly simple in its workings compared to most fish skulls. Even in the Devonian period there were fishes, e.g., *Rhizodopsis*, with skulls made up of large numbers of precisely articulating bony parts that formed a complex mechanical system. I believe that it would be difficult to document objectively the general conclusion that Recent animals are structurally more complex than known Paleozoic members of the same taxa.

Man must, of course, have had morphologically simple metazoan ancestors somewhere in his history, if not in the Devonian period, then before. The question of the relative complexity of man and fish arises in connection with the popular pair of assumptions that (1) evolutionary progress from lower to higher organisms consists of increasing structural complexity; (2) the change from fish to mammal exemplifies such progress. In some respects, such as brain structure, a mammal is certainly more complex than any fish. In other respects, such as integumentary histology, the average fish is much more complex than any mammal. What the verdict after a complete and objective comparison would be is uncertain.

In considering the relative structural complexity of different organisms it is customary to limit discussion to the adults of each type. This is partly justified by the fact that the adult stage is usually the most structurally complex stage in the life cycle, but this limitation also may indicate a relatively naïve view of development. Ontogeny is often intuitively regarded as having one terminal goal, the adult-stage phenotype, but the real goal of development is the same as that of all other adaptations, the continuance of the dependent germ plasm. The visible somatic life cycle is the indispensable machinery by which this goal may be met, and every stage is as rightfully a goal as any other. Each stage has two theoretically separable tasks. First it must deal with the immediate problems of survival, a matter of ecological adjustment. Secondly it must produce the next succeeding stage. The morphogenetic instructions must provide for both jobs. The burden of ecological adaptation must inevitably be heavy in a stage that inhabits a complex and often hostile environment. In stages spent in constant and normally favorable environments, however, very little of the genetic information need be concerned with ecological adjustments, and developmental compromises can be heavily in favor of effective morphogenetic preparation at the expense of the machinery of immediate ecological adjustment. Compare, for example, the kinds of adaptation shown by human foetuses with those shown by children and adults. The foetus lives in an actively cooperative environment and has few ecological problems. It can concentrate on rapid and efficient morphogenetic preparations for later stages. The child or the man lives in a complex and frequently hostile environment. In these stages the emphasis is on precise sensory, motor, immunological, and other ecological adaptations. The morphogenetic preparations are much less fundamental in scope and much slower than those of the foetus.

But suppose the human foetus lived, not in a protective and solicitous uterus, but in an environment like that of a tadpole. Suppose that man's "larval" development, like that of a frog, took place in an environment different from that of the adult and as complex and dangerous as that of an amphibian larva. Man's germ plasm would then undoubtedly be burdened with instructions for coping with conditions as rigorous as those on pond bottoms. Complex sensory and motor mechanisms would be developed early in life, and some of these would need drastic modification for the adult stage.

How much would the total information content of the human zygote be augmented by such additional instructions? There is no answer available at the moment, but if the question could be workably and realistically codified, perhaps a formal analysis would provide some understanding. Such an analysis might take the form of asking whether there is more developmental information in two zygotes, one of which develops into A and the other into B, than in a single zygote which develops first into A and then into B. The single zygote, while developing into A must, as an added duty, preserve the information necessary for producing B, and producing B may necessitate an undoing of part of A. Would this mean that the more complex life cycle would require more information than the two simple ones? Until sound arguments are formulated we must be wary of passing judgments on the relative complexities of organisms of very different life cycles.

There are life cycles enormously more complex than that of a frog. The lowly and "simple" liver fluke develops from a zygote into a multicellular miracidium, which swims by means of a covering of thousands of cilia and has the neuromotor machinery necessary for locating and burrowing into a certain species of snail. Inside a snail it metamorphoses into a morphologically different sporocyst, which reproduces by internal budding. The products of this reproduction are another stage, called the redia. The rediae migrate within the snail and reproduce other rediae asexually. Eventually the rediae metamorphose into another type, the cercariae, which are equipped like the earlier miracidia for migration between hosts, but with quite different motor mechanisms. A cercaria swims by wriggling a tail, not by the action of cilia. A cercaria burrows out of the snail and swims to a blade of grass, to which it attaches, and changes into a dormant, more or less amorphous multicellular mass called a metacercaria. On ingestion by a sheep, a metacercaria hatches out as a young fluke, which develops into an adult fluke within the sheep. The adult flukes produce zygotes which then repeat the cycle. Such a complex succession of morphologically different stages must demand much more in the way of morphogenetic instructions in the germ plasm than would ever be suspected on the basis of the structural complexity visible at any one time. I can see no reliable way at present of evaluating the relative morphological complexity of such different organisms as a sheep and the fluke in its liver, and no way of determining which has the greater burden of morphological instructions or total genetic information in its zygote.

The apparent ease with which trematodes and other parasites can add and subtract distinct morphological stages in their life cycles has a bearing on the problem of the proportion of genetic information that relates to morphogenesis. Perhaps it means that the instructions for producing a cercaria, for instance, are a very trivial part of the total information that must be carried in the fluke zygote.

PROGRESS in animal evolution is sometimes assumed to mean increasing histological differentiation. Such progress, like increasing morphological complexity, must have occurred somewhere in the development of all of the Metazoa. Also, I am inclined to concede that mammalian tissues may be physiologically somewhat more specialized than those of a fish. Such tissue specialization is apparently acquired at the price of regenerative abilities. To a certain extent this implies the substitution of one adaptation for another, not merely additional adaptations. The concept of progress as tissue specialization would probably have little appeal in an application outside the vertebrates. Such cell-constant organisms as rotifers and roundworms would have to be considered higher animals than the mammals. Their tissues are so specialized that they even lack effective mechanisms for the healing of minor wounds (Needham, 1952).

Many of the concepts of evolutionary progress and the implied judgments of the degree of advancement of different organisms adhere to the forms of an earlier orthogenetic doctrine even though the doctrine is almost unanimously discredited. It would be in line with current practice to note that evolution has, in fact, produced an organism of special interest, such as man or the horse. Progress is then arbitrarily designated as any change in the direction of man or horse. As a convention most biologists would accept the judgment that *Pliohippus* is more advanced than

Mesohippus, and that *Australopithecus* is a higher form than *Proconsul*. In groups in which there are no end products of anthropocentric importance, as in the flowering plants and the fishes, there is still a conventional recognition of lower and higher forms. It is observed in the fishes that a number of faunally important groups have independently undergone certain developments: an upward shift of the pectoral fins and a forward shift of the pelvics; the establishment of relatively low and constant numbers of vertebrae, fin rays, and other meristic parts; loss, in the adult, of the embryonic connection between the air bladder and the gut; development of defensive spines in various parts; etc. Groups that show many such developments are considered higher than those that show few or none. The mere extent of departure from primitive conditions is another important consideration. The flatfishes, with their drastic reorganization of the primitive bilateral symmetry and their many other striking modifications, are always accorded a high place by systematic ichthyologists.

Biologists who are especially zealous in avoiding such concepts as evolutionary progress will sometimes use the term *specialized* instead of *advanced*. This term, however, has value in an ecological context that is independent of phylogenetic position. Thus a pike and a bluefish are similarly and perhaps almost equally specialized for a fish diet, but one is low and the other high on the ichthyologist's conventional scale.

There is no objection to the use of such terms as *progress* and *advance* to designate conformity to common phyletic trends or approach to an arbitrarily designated final stage, but unfortunately the accept-

NATURAL SELECTION, ADAPTATION & PROGRESS

ance of the term in this sense can disguise its use in other senses. Thus mammalogists, using extensive and objective evidence, classify the primates into suborders, families, and genera, and then list these categories with tree shrews at the beginning and man at the end. The acceptance of this classification then makes it easy to imply that progress toward man is a recognized evolutionary principle that has operated throughout the history of the primates. I suspect that evolutionary progress and the inevitability of man may seem like scientific ideas only because of our heritage of such orthogenetic terms as "higher" or "advanced" organisms and the fact that a list of taxonomic categories has to have a beginning and an end.

PROGRESS is also commonly taken to imply improvement in the effectiveness of adaptation in a way analogous to technological improvements in man's implements. Huxley's (1954) treatment recognizes such improvement as a common result of evolution but restricts the recognition of progress to a relatively few, especially promising sorts of improvement. Brown (1958) makes a similar distinction between special adaptations and general adaptations. Brown is more liberal in his recognition of general adaptation than Huxley is in the recognition of progress. Unlike Huxley, Brown believes that there is still abundant opportunity for evolutionary progress. Waddington's (1961) concept of progress seems closely related to Brown's and emphasizes independence of environmental change as an important component. To Thoday (1953, 1958) progress means improvement in the long-term effectiveness of adaptation so as to make the population less likely to become extinct. Most of my discussion on progress as improvement of adaptation will apply to Thoday's concept, but the problems of the fitness of a population, as measured by how effectively it avoids extinction, will be considered mainly in later chapters.

It is certainly true that some evolutionary developments, such as the specialization of certain Devonian fishes for a marginal and often anaerobic habitat, can precipitate adaptive radiations of great importance, and that other developments have had no such consequences. Unfortunately, no one has proposed any objective criteria by which we might, *a priori*, distinguish the categories of progressive and restrictive changes. I will confine the present discussion to the treatment of the concept of adaptational improvement as synonymous with progress, or at least an aspect of it.

It may help to start with an analogy in human artifice. We would regard a modern jet-powered aircraft as more advanced than a propeller-driven craft. It should be noted that this need not imply that the improvement involves an increase in complexity. On the contrary, the jet is in basic plan much simpler, but it represents a greater achievement of engineering, and it is in many ways a better engine. The jets have rapidly replaced their ancestors in both military and commercial applications. The propeller-driven craft may not be facing complete extinction, but it has disappeared from many of the fields in which it was once dominant, and it has lost ground in many others.

There seem to be many analogies in organic evolu-

NATURAL SELECTION, ADAPTATION & PROGRESS

tion. The gnathostomes almost entirely replaced the agnaths, presumably because they were more effective fishes. The angiosperms largely replaced the gymnosperms, presumably because they were more effective terrestrial autotrophs. The Carnivora entirely replaced the creodonts, presumably because they were more effective mammalian carnivores; and so on. On the other hand, in the Mesozoic era the newly evolved reptilian mososaurs, plesiosaurs, and ichthyosaurs contested the seas with large carnivorous fishes, such as the ancient sharks. The sharks are still here in great abundance, while their reptilian competitors are all extinct. The Pliocene epoch saw the mass extinction of many of the higher mammalian carnivores, ungulates, and primates, but more primitive mammals and lower groups were little affected. Today the fishery biologists greatly fear such archaic fishes as the bowfin, garpikes, and lamprey, because they are such outstandingly effective competitors and predators of valued teleosts like the black basses and salmonids. I cite these examples not because I believe that the lower forms usually prevail over the supposedly more advanced, but simply to show that the game can be played both ways. The citing of selected examples of the supposed operation of a process, such as the dominance of recently evolved types over the more ancient, cannot be accepted as evidence for the process. Only an unbiased and statistically significant list of examples would be acceptable. Such evidence is conceivably obtainable, but I know of no attempt to obtain it.

I must concede that some of the traditional examples are impressive in themselves. The triumph of the placental mammals over the marsupials in South

America gives every indication that the placentals are, by and large, better adapted. It is tempting to attribute the success of the placentals to such characters as their larger brains and their chorionic placentas, by which we recognize them as more progressive forms. But even here there are other possible interpretations. There is reason to believe that representatives of rich biotas, such as that of the Holarctic, are usually better able to invade new areas than representatives of poor biotas like the Neotropical, regardless of their positions on phylogenetic scales. There may also be a purely statistical factor at work. If there were many more genera and species of placentals in North America than there were marsupials in South America, we would expect many more successful north-to-south migrants through the isthmus than south-to-north ones. Even if it could be demonstrated that the placentals are adaptively superior to the marsupials, this would be only one example of the superiority of a supposedly higher form over a lower.

An even better example is the rapid, worldwide triumph of the angiosperms over all other forms of terrestrial autotrophs. Most of the more philosophical discussions of progress, however, have little to say about the plants. The botanists do make use of the concept of advancement on a phylogenetic scale, with increased specialization to the terrestrial habitat as the main criterion. Such things as the absence of dependence of the fertilization process on an aquatic medium of sperm transport receive particular emphasis. Conformity to general phylogenetic trends, especially in flower structure, is another important consideration. It would certainly be reasonable to attribute the triumph of the angiosperms to the terrestrial specializations seen in the vascular and reproductive systems.

If it should turn out that the weight of evidence favors the conclusion that what we intuitively regard as higher organisms are adaptively superior to the more primitive types, it will obviously be only a minor statistical bias, with many notable exceptions. Despite the supposed inferiority of the adaptations of amphibians—and any general zoology text can supply an impressive list—the modern anurans and urodeles seem abundantly successful. If numbers of individuals or of species is any criterion, as it is often assumed to be, we live in an age of amphibians as much as an age of mammals. The amphibians compete directly with reptiles, birds, and mammals for food and other essentials, and do not seem to be at a great disadvantage. There are many examples of ancient phyla of presumably low development that are abundant in species, individuals, and biomass even though they are often in close competition with presumably more progressive groups. The sponges and hydroids are more in evidence in coastal waters than the bryozoans and ascidians. I can think of no more important evidence on this problem than the obvious fact of the continued success of ancient and supposedly inferior types.

The most apparent explanation is that the taxonomic diversification of life has been mainly a matter of the substitution of one adaptation for another, independently in different lines of descent, rather than an accumulation of adaptation, as would be implied by the term "progress." The original tetrapods became better walkers at the price of becoming inferior swimmers. The original homoiotherms decreased their metabolic dependence on environmental temperatures, but thereby increased their requirements for food, and so on. There were undoubtedly some important, long-term, cumulative trends in the early evolution of life. Some may have continued even after evolution became stylized by the establishment of precise chromosomal inheritance and sexual reproduction. Some may even be in evidence today, and some may be of a kind that would suggest the term "progress." The demonstration and description of such trends are matters of scientific interest and deserve some attention from evolutionary biologists. On the other hand, it seems certain that within any million-year period since the Cambrian such trends were of very minor consequence. The important process in each such period was the maintenance of adaptation in every population. This required constant rectification of the damage caused by mutation, and occasionally involved gene substitutions, usually in response to environmental change. Evolution, with whatever general trends it may have entailed, was a by-product of the maintenance of adaptation. At the end of a million years an organism would almost always be somewhat different in appearance from what it was at the beginning, but in the important respect it would still be exactly the same; it would still show the uniquely biological property of adaptation, and it would still be precisely adjusted to its particular circumstances. I regard it as unfortunate that the theory of natural selection was first developed as an explanation for evolutionary change. It is much more important as an explanation for the maintenance of adaptation.

NATURAL SELECTION, ADAPTATION & PROGRESS

I believe that my point of view on the subject of progress and of changes in the mechanisms of adaptation is really the prevailing one in the laboratory and the field and in the technical literature of biology. It is mainly when biologists become self-consciously philosophical, as they often do when they address nontechnical audiences, that they begin to stress such concepts as evolutionary progress. This situation is unfortunate, because it implies that biology is not being accurately represented to the public.

CHAPTER 3

Natural Selection, Ecology, and Morphogenesis

IN THIS chapter I will present some points of view that may facilitate the search for connections between the natural selection of alternative alleles and the phenomena of ecology and morphogenesis. It is my contention that the production and maintenance of adaptation in these realms is adequately understandable without recourse to creative evolutionary forces that would not be predictable outcomes of selective gene substitution. I will also discuss some examples of the supposed inadequacy of natural selection to resolve certain problems of ecology and morphogenesis.

The relationship of genotype to phenotype is that different genotypes may produce different phenotypes in the same environment. The genotype is a coded message that is interpreted in some way by the soma. A gene is one of a multitude of meiotically dissociable units that make up the genotypic message. No constant phenotypic effect need be associated with a particular gene. The substitution of one allele for another may have one effect in one genotype and an entirely different effect in another, and only the entire message can be said to have a meaning.

Obviously it is unrealistic to believe that a gene actually exists in its own world with no complications other than abstract selection coefficients and mutation rates. The unity of the genotype and the functional subordination of the individual genes to each other and to their surroundings would seem, at first sight, to invalidate the one-locus model of natural selection. Actually these considerations do not bear on the basic postulates of the theory. No matter how functionally dependent a gene may be, and no matter how complicated its interactions with other genes and environmental factors, it must always be true that a given gene substitution will have an arithmetic mean effect on fitness in any population. One allele can always be regarded as having a certain selection coefficient relative to another at the same locus at any given point in time. Such coefficients are numbers that can be treated algebraically, and conclusions inferred for one locus can be iterated over all loci. Adaptation can thus be attributed to the effect of selection acting independently at each locus. Although this theory is conceptually simple and logically complete, it is seldom simple in practice and seldom provides complete answers to biological problems. Not only do gene interactions and the processes of producing phenotypic effects offer a universe of problems for physiological geneticists, but the environment itself is a complex and varying system. Selection coefficients can be expected to change continually in all but the most stable environments, and to do so independently at each locus.

In dealing with the complexities of the selection of alternative alleles it may help to think of the environment as something more extensive than that commonly considered by ecologists. I find it convenient to recognize three major environmental levels, the *genetic*, the *somatic*, and the *ecological*.

THE MOST intimate environment in which a gene is selected is the other genes at the same locus. Gene *a* may be favorably selected in a population in which the normal allele at the *a*-locus is *A*, but unfavorably in a population in which it is mainly *A'*. It might be favorably selected initially, when it is so rare as to be judged only in the combination *Aa*, but if the homozygote is less fit than the heterozygote, the selection coefficient of *a* will drop to that of *A* as it increases in frequency. Problems involving several alleles can be met by a simple logical extension of the diallelic model (Wright, 1931). The mathematical complexities that arise do not impugn the general principle that at any time a given gene will have a certain selection coefficient relative to its alleles, and that this number will determine (aside from statistical error) whether it increases or decreases with the passage of generations.

Genic selection coefficients can also depend on the genes at other loci. Gene *a* may be favorably selected in genotypes *BB* and *Bb*, but unfavorably in genotype *bb*. Its selection coefficient would depend algebraically on the magnitude of the differences and on the relative numbers of the two alleles at the *b*-locus. The genetic environment can be considered to be all the other genes in the population, at the same and other loci. In practice, it is possible to consider only a few of the more important loci and treat the remainder as error or noise. A formal definition of selection coefficient as a function of the genetic environment would be

$$\overline{W} = \Sigma\,(P_i W_i)$$

where W_i is the coefficient in the ith genetic environment, and P_i is the relative frequency of that environment. The selection coefficient of the gene a as a function of the environment at its own locus and one other, with two alleles at each locus, would be

$$\overline{W}_a = P_{ABB} W_{ABB} + P_{ABb} W_{ABb} + \ldots + P_{abb} W_{abb}$$

Any change in the genetic environment would be represented by complementary changes in at least two p-values, and this would influence the value of the selection of coefficient $\overline{W}_a$. In practice, one normally calculates selection coefficients by observing changes in gene frequencies and departures from Hardy-Weinberg equilibrium values, and does not derive them from such components as are used in the equations.

The term *genetic environment* was introduced by Mayr (1954) and is of value for its emphasis on the genetic composition of the population as an aspect of the environment in which the selection of a gene takes place. I am unaware of any use of the term subsequent to Mayr's, but the concept is certainly widely appreciated and understood. It is often referred to as the genetic background, and it is implied in discussions of the *integration* of the gene pool or of *coadaptation* among genes. Levene, Pavlovsky, and Dobzhansky (1958) provided a good illustration of the effect of variation in the genetic environment. They showed that the outcome of competition experiments with two karyotypes depended on the stocks used to carry the chromosomal types under consideration, in other words, on the genetic environments in which the chromosomes were placed. A theoretical analysis of selection in a two-locus environment was contributed by Lewontin and Kojima (1960).

Genotype frequencies would be more or less strictly determined by gene frequencies in any large population. Even where balanced polymorphisms or other factors cause serious departures from Hardy-Weinberg distributions, the departures themselves would be determined by gene frequencies and the effects of different genetic environments on selection coefficients. It is therefore legitimate to think of a gene pool as a single genetic environment in which a given gene substitution must be judged. Each allele will have its own particular selection coefficient in a particular gene pool. The fact that this genetic environment is really made up of an astronomical number of genetic subenvironments, in each of which the gene may have a different selection coefficient, can be ignored at the level of general theory. At times it may be necessary to investigate selection in the different genetic environments of one population for an understanding of a particular problem of adaptation. In very malarial parts of Africa a common gene produces homozygotes with a fatal disease, sickle-cell anemia, and viable heterozygotes with high resistance to malaria (Allison, 1955). Its allele produces a homozygote that is normally viable but normally susceptible to malaria. If the gene associated with anemia and malarial resistance is designated S, its selection coefficient in the genetic environment S would be very different from its coefficient in environment S'. Its effective (mean) coefficient would be the mean for these two environments, weighted by the frequencies of the environments. These coefficients change in time and space as a function of the incidence of malaria.

The multiplicity and complexity of genetic interactions are often such as to approach the fluid concept of heredity, rather than the particulate Mendelian concept. Striking differences in selection coefficients as a result of simple differences in genetic environment, as exemplified by the sickle-cell gene, are relatively rare. In most instances one can ignore the problem of discrete genotypes and assume that a gene has a continuous spectrum of expressivity, not only in the usual quantitative sense on conspicuous unit characters, but also in a qualitative manner (pleiotropy) on a variety of characters. In this variety there will often be both positive and negative contributions to fitness, and selection coefficients will reflect the balance of these contributions.

I would conclude that the population gene pool constitutes one aspect of the environment in which the natural selection of alternative alleles must take place. The recognition of this principle is perfectly consistent with the theory that the natural selection of alternative alleles is the only force responsible for the production and maintenance of adaptation.

THE SOMATIC environment is produced by the interaction of the genetic and the ecological and is therefore a sort of intermediate level, but it is convenient in this discussion to treat it separately. I used the term once before (Williams, 1957). The distinction between somatic and genetic may be arbitrary at times. It is customary to think of the egg cytoplasm as the soma of a zygote, but there may be elements in this cytoplasm that are more or less a part of the genetic message. Often the information so transmitted is physically particulate, and the term *plasmagene* is appropriately descriptive. Sometimes there is no evidence on the physical nature of the cytoplasmic influences, although there may be abundant reason for recognizing the cytoplasm as an important developmental variable. For example, Fowler (1961) has shown that a genome that produces normal development in one subspecies of the frog *Rana pipiens* can lead to gross abnormalities if put into the egg of another species. The same nuclear message is interpreted by one soma in one way and by another soma in an entirely different way. Even for an abnormal development a foreign nucleus must be from another individual of the same species or from a closely related form. The genetic message from a mammalian nucleus would be unintelligible to an avian cytoplasm.

The changing meaning of the same genetic message can be seen clearly in the normal development of a multicellular organism. For a time at least, all nuclei of an embryo may be equivalent to each other and to the original zygote nucleus. The equivalence of all, or at least many, nuclei in a higher plant is attested by the ability of widely different tissues to give rise to flowering branches or to structures passed on in asexual reproduction. In the vertebrates the immunological similarities of widely different tissues of the same individual can be regarded, as was pointed out by Michie (1958), as indicating the genetic equivalence of such somatic tissues. The morphogenetic interpretation of the same genotype varies enormously, however, according to the stage of development and position within the embryo. At first the developmental meaning of the genetic message is little more than "perform mitotic divisions" to all parts of a vertebrate soma. A little later the same message means "invaginate" in one place and "just keep dividing" in another. Later, to a cell in one part of the embryo the message means "elongate" and to another cell elsewhere it means "flatten out," and so on. The equivalence of nuclei leaves the epigenetic theory of development as the only possible interpretation. If nuclei in different animal tissues do become genetically different, as is indicated by some of the evidence, such differences must themselves arise epigenetically by nuclear reactions to differing somatic environments.

Among plants there are developmental differences that were originally assigned a genetic basis but which have since been found to result from variations in the somatic environment. For example, the differences between gametophyte and sporophyte were generally assumed to result from the gametophyte development's being normally directed by haploid nuclei and sporophyte development by diploid nuclei. However, it has been shown (Wardlaw, 1955) that gametophyte development can proceed with diploid nuclei, and sporophyte development with haploid, in a wide variety of plants from fungi to spermatophytes. The important factor is the initial somatic environment. A normal nucleus, whether haploid or diploid, will direct the development of a gametophyte if it is in the somatic environment provided by a spore, but a sporophyte if it is in that provided by a zygote. Apparently the genetic message is much the same, whether the nucleus is diploid or haploid, but the interpretation of that message by a spore soma is entirely different from the interpretation by a zygote soma.

If the same genetic message is interpreted by different somata in different ways, it must require different messages to get different somata to produce the same effects. Fowler's work, cited previously, showed that to get normally proportioned frogs from both northern and southern eggs, it was necessary that the nuclei be different, not similar. Genetically similar nuclei in both would produce normal development in one and abnormal development in the other. The close external similarities shown by northern and southern representatives of *Rana pipiens* are due in part to their being genetically different. The production of similarities in such different environments as the cytoplasm of a mammalian egg and a bird egg must necessitate very different genetic instructions. There is an analogy in human communication. If a message to someone who understands only Chinese produced the same response as a message to someone who understands only Japanese, we can be sure that the messages must have been different.

If would follow as a general conclusion from the epigenetic theory of development that whenever morphogenetic processes produce closely similar results in widely different somatic environments, the similarities must depend on genetic differences of a kind and degree that balance the environmental differences. White and Andrew (1962) have reasoned that few if any of the potential "loci" in two mutually inverted chromosomal segments would still be identical after 10^5 generations. Yet such inversions may still be producing phenotypes that cannot be distinguished. Darlington (1958) and Dobzhansky (1959) have advanced other reasons for believing that an evolutionary stability of the phenotype need not imply a comparable stability of the genotype, and that a gene pool may be in a greater state of flux than would be indicated by phenotypic evolution. These conclusions are in opposition to that of Emerson (1960), who argued that the existence of structural homology between widely different taxonomic groups indicates that the germ plasm is a conservative entity and that the homologies are due to the presence of identical elements in the different germ plasms.

The normal succession of somatic stages in which a gene expresses itself is one aspect of the total environment which assigns relative selection coefficients to each of the alleles at every locus. To be favorably selected, a gene need not produce greater fitness than its alleles in every stage of this succession. Its selection will ultimately depend on its mean effect at different stages, weighted by the frequency and duration of each stage. The weight assigned to the stage "first decade of the second century" for man would be zero or very close to it because of its infrequency in the life cycle. The two-cell stage would get very little weight because of its brevity. The importance of such factors as the durations of morphogenetic stages will be considered later. For the present the important point is to recognize that the somatic environments, both in their physical natures and their durations, play an essential role in determining selection coefficients. The recognition of this fact in no way compromises the principle of selective gene substitution as the sole and ultimate force of adaptive evolution.

NATURAL SELECTION, ECOLOGY & MORPHOGENESIS

THE *ecological environment* is the familiar world of the ecologist and the concept usually meant by the unqualified term *environment*. Certain aspects of the ecological environment, such as climate, predators, parasites, food resources, competitors for such resources, etc., are well understood as evolutionary factors. Their treatment receives a large share of attention in the literature on adaptive evolution. It is in relation to these factors that the adequacy of selective gene substitution as the force of evolutionary adaptation is most generally recognized. I will concentrate here on certain problems of ecological adaptation that seem to have had less than a fair share of attention. One such problem is the role of the ecological environment in morphogenesis. Another is a part of the ecological realm that might be called the *social environment*. It is made up of all other contemporary members of the same population—individuals that may supply important resources, may be ecological competitors, and are always genetic competitors. A consideration of the social environment is reserved primarily for Chapters 6 and 7. Another ecological subcategory might be termed the *demographic environment*.[1] It includes the inevitable her-

[1] The study of population size and density, the ratios of age groups or other classes of individuals, and the rates of such events as birth and death involves much the same body of concepts and problems whether the data derive from human, biological, or hypothetical populations. This broad field of study should have a name, and *demography* is appropriate. This inclusive usage is urged by Cole (1957) and implied in many recent biological writings. I will use *demographic* to refer to such measures as crude or age-specific mortality rate, sex ratio, vagility, etc., in both human and non-human populations.

itage of age-dependent probability distributions of reproduction and of death, and will be considered briefly in this chapter. It is in relation to these ecological factors that the importance and adequacy of the theory of genic selection is least understood.

The boundary between the somatic and ecological environments is not entirely distinct. Sometimes a major phenotypic difference may result from a minor and transient ecological factor, largely by a triggering of a critical change in the somatic environment. A minor dietary change early in development makes a bee a worker rather than a queen. The meaning of the genetic message is different in a worker soma from what it would be in a queen soma. The presence or absence of metamorphosis in urodeles may be similarly decided by a dietary threshold. The settling and metamorphosis of marine larval stages is contingent upon sensory stimuli emanating from suitable attachment sites. Plants subject to certain periodicities of darkness and illumination may become determined to flower at some later time, even if the external periodicity is changed. In these well-known examples the ecological environment decides at one point which of two alternative morphogenetic developments will be realized, and thereafter the somatic environment is the governing influence.

It is the ecological environment that determines how well adapted a soma will be and what sorts of morphogenetic change are possible. We can regard the ecological environment as the strategy employed by Nature against an organism which, in turn, replies with a strategy of its own that is designed to win the highest probable score (number of successful offspring). The ecological environment has a strategy in the formal sense of game theory. Any system of ploys is a strategy by definition. Nature, however, is by convention assumed to have no strategy at all in the vernacular sense, but to play at random, that is, without regard for whether she wins or loses. In game theory, therefore, an organism's ecological environment is not Nature. The environmental ploys are a more effective strategy than would ever be produced at random. This is so because the ecological environment is full of other organisms that have effective strategies of their own. At a general level, these various strategies are independent. The goal of the fox is to contribute as heavily as possible to the next generation of a fox population. The goal of a rabbit is to do the same in a rabbit population. Neither uses a strategy specifically calculated to frustrate the other. However, the achievement of the fox's goal may require, at the tactical level, the death of the rabbit, an event inimical to the rabbit's strategic interests. At the tactical level such organisms will often operate at cross-purposes, even though their long-term strategies are independent. As soon as an organism's defense mechanisms are greatly reduced in effectiveness, as they always are by death, it is rapidly destroyed in the biosphere. Only a lifeless Earth would really behave as Nature is supposed to in game theory, and a dead body in such a world would last for an immensely long time.

The effectiveness of the environmental strategy arises only as an incidental result of the strategies of other organisms. The ecological environment seeks no saddle-points in its game with an organism. Its strategy is highly imperfect, and it makes many ploys that either benefit the organism directly or can be made to do so, by the organism's suitably adjusting its own strategy. Thus the ecological environment provides sunlight for the grasses, and rabbits for the foxes, and water for both. The bearing of these matters on such concepts as the organization of the community will be discussed on pp. 246-250.

The ecological environment provides resources, not only in the form of food and other requisites for survival, but also in the form of contributions to morphogenetic processes. One way in which this occurs is by the organism's choosing the ideal niche within its general environment. Choice may be mediated by the central nervous system, in a manner acceptable to the vernacular use of the word *choice*. Thus it is undoubtedly due to something in the neural and psychic activities of its brain that a spider monkey spends most of its time in the trees. In the theoretically important sense, however, the concept of choice is a broader one. In the open sea, a duck floats largely above the surface of the waves and the tuna swims beneath them. This may in part be due to the psychically determined seeking of an aerial environment by the duck and a submerged one by the tuna, but the psychic factor is clearly not the whole explanation. A more important factor is that, by directing the construction of a soma with air-sacs inside and airy hydrophobe feathers outside, the duck genes have chosen a life spent largely above the surface of the sea. Similarly the tuna genes, by specifying an animal of about the same specific gravity as sea water, have cast their fortune in submerged habitats. For both, the chosen environments exert influences on development, and these influences are normally favorable. The duck and the tuna can thrive and mature in their chosen subenvironments of the pelagic realm, but neither could long survive in that of the other. Since the ecological environment is so largely chosen by the soma during development, and since the choice is of precisely the best environment available, we can say that the selection of an ecological environment is part of the normal machinery of development. The availability of atmospheric oxygen is just as genetically determined as the production of mitochondria. Both result from the interaction of the genotype with its various environments. Development cannot be regarded as a self-contained package of activity but as a program of events in which selected parts of the ecological environment form specific components of the epigenetic system. In more than a rhetorical sense, the organism and environment are parts of an integrated whole. The "fitness of the environment" is a reality, but only because the organism chooses its own effective environment from a broad spectrum of possibilities. That choice is precisely calculated to enhance the reproductive prospects of the underlying genes. The succession of somatic machinery and selected niches are tools and tactics for the strategy of the genes.

A habitat is fit because its occupant provides itself with a near-optimum soma for life in that particular habitat. The precision of this adaptation may be compromised by its future commitments. At any stage in its life history, an organism must not only adapt to its immediate circumstances, it must retain the ability to adapt to those likely to be encountered in the future. This requirement implies not only that it retain the necessary genetic information, but that it use only those somatic structures that will be adaptive

NATURAL SELECTION, ECOLOGY & MORPHOGENESIS

in the future as well as the present or can be modified for the future with a minimum of reorganization. A complex multicellular organism cannot be an autotroph one day and a predator the next. The succession of somata in the life cycle of an organism must provide an adjustment of each stage to the one before and the one after, in addition to an adaptive selection of environmental niches and precise somatic adaptation to each niche.

Thus the phenomenon of fitness can be seen at all epigenetic levels, from genic interactions to the ecological niche. It also has extension in time from microsecond events at the molecular level to successions of stages in the life cycle and adjustments to the diurnal and seasonal cycles. All of this is a logically inevitable result of the natural selection of alternative alleles in Mendelian populations. Each level offers innumerable problems for our understanding of adaptation and each is a legitimate field of investigation. It is at the level of the gene, however, that we have the most fundamental and most universally applicable understanding of adaptation.

THE MOST prominent recent challenge to the adequacy of natural selection for morphogenetic phenomena is that propounded by Waddington (1956 *et seq.*), who argued that natural selection must be supplemented by another process, which he calls *genetic assimilation*. His conviction that natural selection is not enough is apparent from his statement that the theory of genetic assimilation "goes some way—though by no means the whole way—towards filling the major gap in Darwin's theory of evolution" (1958, p. 18); and from his statement that with genetic assimilation "we can reduce our dependence on the abstract principle that natural selection can engender states of high improbability" (1959, p. 398). It is easy, on superficial acquaintance, to overestimate Waddington's departure from current tradition and even to regard it as Lamarckian. It is also easy to make the opposite mistake and regard his conclusions as entirely compatible with the traditional model of natural selection. I will therefore discuss Waddington's views of adaptive evolution in some detail, so that I can point out exactly where I find them unacceptable.

The phenomenon of genetic assimilation is a real one, and throws important light on the nature of the genetic control of development. The best experimental demonstration of the phenomenon is Waddington's work on the assimilation of the *bithorax* phenotype. He subjected some fruit-fly eggs to ether vapor in sublethal doses. Most of the survivors gave rise to normal flies, but a few developed into an abnormal type, called bithorax. These abnormal flies were selected as the parents of the next generation and the ether treatment repeated on their eggs. This treatment with ether and selection of bithorax was continued for many generations. The incidence of bithorax in this selected line increased steadily. The most significant observation was that after a number of generations, some eggs in the selected line produced the bithorax condition even without exposure to the ether. Within thirty generations of the original selection, Waddington produced a stock in which a large proportion of bithorax flies appeared every generation, even without the ether treatment. Bithorax was at first an individually acquired character, de-

NATURAL SELECTION, ECOLOGY & MORPHOGENESIS

veloped as a result of an environmental influence on development. At the end of the experiment it had become, in the selected line, a hereditary character. All relevant experimental controls, such as replications and parallel lines in which bithorax was selected against, were performed. Other characters were assimilated in other experiments. There can be no doubt about the reliability of these observations.

Waddington's interpretation of the initial appearance of bithorax may be summarized as follows. First of all, there is a certain specificity in the original stimulus. Specific environmental stresses on development, for instance ether, tend to produce specific abnormalities, such as bithorax. Secondly, any strongly abnormal environmental factor, such as a near-lethal dose of ether, will greatly increase the variability of development. A normal fruit-fly genotype is the result of selection for the reliable and precise production of a certain normal phenotype in the particular range of environmental conditions usually encountered by the species. This genotype is not designed to produce the normal phenotype in an environment that contains a high concentration of ether fumes. Hence, a general increase in variability may be expected in such abnormal environments. Thirdly, there is much unexpressed genotypic variability in the original population. There are a number of genes present, at different loci and in different individuals, that have some tendency to produce the bithorax phenotype. These variations, like normal environmental variations, have their effects suppressed by the normal self-regulatory developmental processes, which Waddington calls *canalization*. However, this genetic variation can become expressed as a result of the ether treatment. The specificities of this treatment, plus the augmentation of variability (decrease of canalization) by the ether, result in the production of a weakly developed bithorax condition in those individuals that have the strongest genetic tendencies in that direction. The continued selection of such individuals results in a rapid increase in the concentration of genes that favor the production of bithorax. Eventually, such genes become so numerous in each individual that they can, by their combined effect, match or exceed the bithorax-producing effect of the ether on a normal genotype. When the stock has been changed to this extent, the ether becomes unnecessary.

This explanation is not in any way Lamarckian. Selection of chance differences between individuals was the evolutionary force that produced the bithorax stock from the normal stock. The environment, however, played a role that is not recognized in the traditional model of natural selection. The ether did not, in the Lamarckian sense, produce the genetic variation that was selected by the experimenter, but it certainly did produce the expression of that variation. Without this expression, there could be no selection and no production of a genetically bithorax stock.

On the basis of such experimental results, Waddington envisioned a role for genetic assimilation in evolution, and he argued that this process provides a mechanism whereby populations can respond very rapidly to changed conditions. On the ordinary view, selection would act on genetic variation in preexisting characters when the environment changed, or, if the necessary character did not yet exist, it would have

NATURAL SELECTION, ECOLOGY & MORPHOGENESIS

to wait for new mutations to fill the deficiency. In the experiment described there was no preexisting character of bithorax, nor, probably, were there any important mutations during the period of selection. Yet a major evolutionary change took place with extreme rapidity by genetic assimilation.

The experiment is of great importance for its demonstration of a previously unsuspected store of latent genetic variability, but I question its value as a model of adaptive evolution. One source of difficulty is in Waddington's tendency to think of the development of bithorax, after an ether treatment, as a response to a stimulus. The term "response" usually connotes an adaptive adjustment of some sort, and would not be used for disruptive effects. It would be normal usage to say that some Frenchmen responded to the Terror by conforming to the Jacobin demands and that others responded by fleeing the country. It would not be normal usage to say that some responded by losing their heads. Decapitation was the result, not of a response, but of a failure to respond soon enough or in an effective manner. Similarly we may say that some of the flies responded adequately to the ether and produced a normal phenotype in spite of the difficulty presented by the treatment. Others showed an inadequate response. They were able to survive in the protected confines of the culture bottle but could only produce a grossly imperfect phenotype, bithorax. By favorably selecting the bithorax condition, Waddington produced an extreme but simple kind of degenerative evolution. He was selecting for specific kinds of inadequacies in the mechanisms of developmental canalization. I would suspect that there is less genetic information in the bithorax stock, which was produced by selection, than there was in the original stock.

Waddington apparently sees no need to distinguish between response to environmental stimuli and susceptibility to environmental interference. It is my belief that these classes of phenomena are utter opposites and that no more fundamental distinction can be made. Here in brief is my dispute with Waddington.

It is possible to confuse responses and susceptibilities because both conform to a pattern of cause-effect relationships involving organisms and their environment. It is important that they not be confused because a response shows the unique biological property of adaptive organization, and susceptibility results from the absence or deficiency of this property. For a response to occur, there must be sensory mechanisms that perceive particular aspects of the environmental situation and activate effectors that efficiently prevent the same, or other, correlated environmental factors from producing a certain undesirable effect. Susceptibility results from the environmental factor's getting through and producing an effect in spite of any responses that may be activated.

The distinction is important enough to warrant an illustrative hypothetical example. Suppose we were to attach to each of two experimental animals—a man and a large reptile—a pair of physiological monitoring devices, one that records the moistness of the skin and another that records heartbeat. Next we put the two organisms in a room where the temperature oscillates slowly between 20° and 40°C. We would find that a physiological measurement records, for both organisms, a history of temperature fluctuations, but that a different physiological variable makes the record in each case. For the reptile we could establish a simple mathematical relationship between temperature and heartbeat, but we would find no record of variation in skin moisture. In the man we would find that skin moisture gives a reliable indication of environmental temperature, at least within much of the studied range, but that little could be learned about temperature from heart rate. The explanation for the difference illustrates the present point. Temperature fluctuations get through to the reptile and produce a direct effect. The temperature of the heart and the rest of the body follow that of the environment and, other things being equal, heart rate is a function of heart temperature for purely physical reasons. By contrast, the temperature fluctuations do not get through to the man. They are perceived by sensors that activate special effectors (sweat glands) to a degree adaptively appropriate to both the ambient temperature and the activity of the organism. The sweat glands and other mechanisms of temperature regulation prevent the environmental changes from having an effect on heartbeat.

The reptilian variation in heartbeat, and the human variation in skin moisture are both *effects* of a certain *cause*, but a biologist should regard them as examples of very different kinds of cause-effect relationships. Waddington fails to make this distinction, and uses the evolutionary origin of a susceptibility to illustrate the origin of a response.

It might be argued that the evolutionary relevance of the bithorax experiment in no way depends on the supposition that bithorax is adaptive or should be called a response. The great majority of phenotypic abnormalities that would result from the ether or other extreme treatments would be expected to be nonadaptive. We likewise believe that the majority of mutations are harmful, but this does not prevent us from believing in mutation as the basic source of variation in evolutionary change. This argument would be valid if genetic assimilation were thought to operate only on very slight changes. The theory of the "hopeful monster" of the early mutationist school has presumably been discredited, and I feel that the arguments against hopeful monsters are equally valid whether the monstrosities are genetic or epigenetic in origin. On the other hand one might say that the bithorax experiment is merely an extreme example of the generally acknowledged principle that environmental changes can alter the expression of genes. The experiment would be considered a macroscopic model of a process that, on a microscopic scale, would be of evolutionary importance. This would seem to violate the spirit of Waddington's proposal, because he believes genetic assimilation to be of primary importance in providing for the rapid development of really novel adaptations.

There are any number of individually acquired characters that are obviously adaptations, and not merely disruptions of development like bithorax. Might not some of these be genetically assimilated and play an important role in evolution? The best examples are inferential rather than experimental and comprise what are called pseudoexogenous adaptations. For example, wherever the human skin is subjected to frequent friction it becomes thicker and tougher and forms a callus. The sole of the foot is the region most subject to friction, and, appropriately, it develops the most pronounced callus layers. This seems like a simple example of an adaptive individual response, but it happens that the thickening of the soles relative to the rest of the skin is a process that starts *in utero*, before any possible frictional stimulus. It would seem that what is normally, for the rest of the body, an individual response, has become a genetically fixed adaptation of the foot. This certainly has the look of a response that has become, in part, genetically assimilated.

Somewhere in man's ancestry there may have been a protoamphibian that occasionally came out on land and pushed itself along with its fin lobes. Such animals might have responded by a minor thickening of the part of the skin that came in contact with the ground, just as our skins thicken wherever they are subjected to friction. At this earliest stage we are dealing with what is entirely an acquired character. Among those that acquired the character, not all would be genetically identical in their ability to develop calloused "soles." If terrestrial locomotion became an important capacity, and if the development of calluses on the feet were an important component of this capacity, there would be selection in favor of those individuals best able to develop this character. In the selected line, those genes that promote the sole thickening would become more and more concentrated, and ultimately there would be individuals that would develop the response purely on the basis of these genetic tendencies, and without any assistance from what had been the necessary stimulus. An individually acquired adaptation of the fishes would have become an obligate adaptation of the tetrapods.

This adaptation, with genetic variation in its development, would have arisen at the same time as the ecological demand to which it was related, but not before. There was no preexisting character of sole thickening in the fishes, and no need to wait for the appearance of new mutations to give the character a start. The parallel with the bithorax experiment is obvious.

It seems to me that the sort of process pictured here must have occurred many times, but I would question its importance as an explanation of adaptive evolution. To explain adaptation by starting with a facultative response and ending with an obligate response is to beg the question entirely. The process starts with a germ plasm that says: "Thicken the sole if it is mechanically stimulated; do not thicken it if this stimulus is absent," and ends with one that says: "Thicken the soles." I fail to see how anyone could regard this as the origin of an evolutionary adaptation. It represents merely a degeneration of a part of an original adaptation. If the origin of the sole thickening as a fixed response is hard to explain, surely its origin as a facultative response is much more so. It must, as a general rule, take more information to specify a facultative adaptation than a fixed one.

Thus at the most general theoretical level, all of Waddington's examples of genetic assimilation would be cases of degeneration, not adaptive evolution. This need not mean that all facultative responses represent higher levels of adaptation than all fixed responses. The confinement of the obligate calluses to particular parts of the body, and the specific patterns that they take must require a considerable amount of genetic information. Nevertheless, as a general class, facultative adaptations represent more difficult evolutionary attainments than obligate adaptations. To use the facultative as an axiom in explaining the obligate is to turn the whole problem upside down. Warburton (1955) expressed this objection in a forceful manner when he said that to acquire an adaptation in Waddington's sense is like "being sewed into one's winter underwear." Underwood (1954) has also expressed opinions similar to mine on the relation between facultative and obligate responses. A formal theory of the optimization of morphogenetic responses in varying environments is provided by Kimura (1960).

It must be understood that calling a character fixed or obligate does not imply that it is inevitable or invariable. All vital functions are susceptible to environmental interference, given sufficiently great stresses. Likewise there is great variation in the range of possible adjustment in facultative responses. In general the adaptive adjustments would be most apparent in the ecologically normal range of stimuli. Susceptibility to interference would be most common and most marked for uncommon or abnormally severe stresses.

The origin of a fixed adaptation is simple. The population merely needs to have or to acquire some genetic variation in the right general direction. The origin of a facultative response is a problem of much greater magnitude. Such an adaptation implies the possession of instructions for two or more alternative somatic states or at least for adaptively controlled variability of expression. It also implies sensing and control mechanisms whereby the nature of the response can be adaptively adjusted to the ecological environment. A facultative response would require much more delicate genotypic adjustments than a comparable fixed response. As an example from man, I would imagine that the obligate difference in skin color between a light and a dark race could be easily evolved on the basis of a wide variety of possible gene differences, and that such racial divergence could take place rapidly on the basis of whatever genes happen to be available. By contrast, the capacity, found in all races, to adjust the melanin content of the skin in response to variations in solar irradiation is an adaptation that must have taken a much longer time to evolve, and must require the carrying of a much larger burden of genetic information.

Waddington gives very little attention to the origin of the facultative responses with which he starts his arguments. He summarized his attitude in one discussion (1958, p. 17) by postulating that natural selection "would, in fact, build in to the developmental system a tendency to be easily modified in directions which are useful in dealing with environmental stresses and to be more difficult to divert into useless or harmful paths." It would appear that he finds the theory of natural selection entirely adequate to explain facultative adaptations, but feels that this theory has a "major gap" in its application to fixed adaptations.

The principle of the economy of information, discussed in connection with the genetic code in Chapter 2, may be useful in predicting whether an adaptation will be found to be obligate or facultative. Whenever a given character would be more or less universally adaptive in a population, it can be expected to be obligate. Only when adaptive adjustment to uncertain conditions would be important would one expect facultative control. Since the obligate is more economical of information, it can always be expected in situations in which a facultative response would not be significantly more effective. This principle is relevant to the current controversy between the nativists and the empiricists on the use of sensory experience for environmental interpretations. It would be expected that whenever a certain physiological state, for instance the parallel orientation of the optical axes of the eyes when focused on an object, *always* indicates a certain environmental state, such as the great distance of the object, the response, here the interpretation of distance, will be instinctive. Similarly in any animal, man included, for which fear of the edges of precipices would be universally adaptive, we can expect such fear to be instinctive, rather than learned. Complex systems of behavior, such as the more elaborate reproductive patterns, will usually be a blend of learned and instinctive elements. There are things that have to be learned, such as the individual characteristics of a particular mate or the location of a nest site. All elements that can be instinctive, however, will be instinctive. Instinct costs less than learned behavior, in the currency of genetic information.

DARWIN (1882, Chap. 27) observed that the parts of higher animals that are most readily regenerated are those that are most likely to be lost, such as the chelipeds of fiddler crabs and the tails of lizards, and that a high development of regenerative abilities is associated with early stages of individual development and low positions on the phylogenetic scale. Flatworms regenerate better than vertebrates, and vertebrates better than roundworms. Unfortunately our understanding of the evolutionary and ecological significance of regeneration has advanced very little beyond this time-honored foundation. This conclusion is apparent in recent summary treatments of the subject (Needham, 1952; Vorontsova and Liosner, 1960).

Despite the lack of progress in relating regeneration to evolutionary theory, and the occasional statements that this theory is inadequate in the treatment of regeneration (see below), I would suggest that genic selection and the concomitant principle of economy of information can be potent allies in any attempt to explain the phylogenetic variation in the phenomena of regeneration. The elaborateness and precision of morphogenetic machinery requires a certain burden of genetic information. The extent of this information will be curtailed, and the dependent machinery will degenerate, to the degree that selection for their maintenance is relaxed. Such a saving in usage of genetic code can presumably be utilized in other ways. As a multicellular animal passes through the developmental program, its ability to repeat earlier stages declines in relation to the likelihood of its being called upon to repeat them. A human zygote has both the genetic information and the somatic machinery necessary for producing a right arm. Once the primordium of this limb is formed, however, the embryo will never again be in a position in which it would be an advantage to grow another such arm. The ability would not disappear instantly. Selection would favor a safety factor that would ensure that the ability

NATURAL SELECTION, ECOLOGY & MORPHOGENESIS

would be retained *at least* until needed, and this would normally result in the retention of the ability for a somewhat longer period than needed. For this reason the amputation of a limb bud from an embryo will often be followed by its regeneration. Once the relatively brief safety margin has been passed, intrauterine traumata that would result in amputation would be so highly improbable that there would be virtually no selection in favor of any prolonged period of ability to regenerate the missing member. Very shortly, development will have proceeded to the point at which such amputation would result in a fatal hemorrhage that would remove all possibility of the exercise of regenerative potential.

In another organism, however, such as a fish or a urodele, the amputation of the homologous part would not necessarily be fatal. In such organisms the perfection of the regenerative machinery would be a measure of the relative frequency of the amputation and the ecological importance of the part in question. This simple consideration of the extent of the advantage and frequency of opportunity of regeneration, plus the principle of economy of information, helps to explain much of the phylogenetic and ontogenetic distribution of regenerative abilities.

One of the classic explanations for the degeneration of useless organs is that material that would be used in their construction can be saved and put to better uses. The degeneration of an abstract capacity, which would be used only when it was, in fact, adaptive, would not save any material. If the loss of the ability to regenerate provides any saving, it must be of informational resources. As was argued in Chapter 2, there is reason for thinking of genetic information as a necessarily limited commodity that must be put to the most effective possible use.

It has been argued that certain kinds of regeneration constitute adaptation to circumstances that have never before arisen, and could therefore not have been produced by natural selection. How frequently in the history of a rat population has partial amputation of the liver occurred, and even if it did occur, would it not always result in fatal hemorrhage from the hepatic arteries and portal vein? Yet rats rapidly regenerate missing pieces of liver. Russell (1945) makes much of this supposed inadequacy of natural selection, and even cites the ability of caddis fly larvae to repair historically unique kinds of damage to their enclosing cases. He also cites the ability of vertebrates to produce antibodies to historically new kinds of antigens. Huxley (1942) maintained that hermit crabs may often have their chelipeds bitten off and that therefore selection would favor the ability to regenerate them, but that their additional ability to regenerate the appendages of the abdomen must be ascribed to some sort of absolute adaptation that exists prior to and independent of natural selection. Such arguments indicate a lack of perspective and imagination.

To expect adaptations to correspond precisely in every detail to the historical circumstances that favored their development is unrealistic. A precise adaptation might require more genetic information than one that would give a blanket coverage to a broad category of ecological demand. Amputation, by aseptic and hemostated surgery, of large masses of liver may be a brand new trauma to the rat, but the "amputation" of comparable masses by various kinds of hepatitis must have been going on for millions of years. Such natural amputation may be extensive; it is too gradual to cause massive internal bleeding; and it is as aseptic as any surgery, apart from the pathogen itself. It is quite understandable that those adaptations designed to repair the effects of hepatitis would be triggered by partial surgical hepatectomy. Similarly the regenerative adaptations described by Russell for caddis flies must be ascribed to a generalized ability to repair the case. A *fleur-de-lis*-shaped excision may be called historically unique, but to the caddis fly it is just another hole. It is also understandable that selection for the ability to regenerate a certain part, such as a cheliped, would produce, as a side effect, some degree of ability to regenerate serially homologous structures. Gene mutations commonly have parallel effects on serially homologous structures, and their selective accumulation would produce a cumulative parallelism of effect.

ONE of the most frequently used examples of the way chance, rather than fitness, may determine survival is the engulfing of large masses of plankton by whales. The supposition is that while whales may be an important source of mortality to a euphausid population they play little or no role in the natural selection of the euphausids. This argument neglects an important aspect of fitness, rapidity of development. If two euphausids are of the same age, but one has already reproduced and the other has not, a whale can exert a powerful selective influence by swallowing both at the same time. Natural selection will always, *ceteris paribus*, favor rapid development; the sooner an organism matures the less likely it is to die before maturing and reproducing. Selection can never directly favor a lengthening of the juvenile period. The development of longer juvenile phases in a phyletic line must always be considered a price paid for some more important development. This principle illustrates the importance of the demographic environment as an aspect of natural selection. When a zygote enters a population it acquires a probability distribution of mortality imposed on it by the ecological environment. It can be expected to be adapted to this circumstance as well as to any other.

Fisher (1930) observed that the immaturity of theoretical biology is attested by the way its practitioners confine their attention to the demonstrably possible. They seldom trouble themselves to explore the much larger realm of the merely conceivable. Fisher illustrated his position with the problem of why there should so generally be two sexes. He maintained that a full understanding of the problem could only be achieved by a rigorous investigation of the consequences of there being more than two sexes. An equally good illustration would be the problem of understanding actual rates of development. Our understanding can be enhanced by considering some patently impossible but conceptually tractable rates.

For instance, suppose there were an organism that took one year to mature, and during this interval had a probability of death of 0.5 per *week*. We would have to endow this species with the astronomical fecundity of about 10^{15} zygotes per pair to keep it from rapidly becoming extinct. Even so, the situation would be extremely unstable. In this species a mutation that accelerated development by only 2 per cent would almost double the probability of its reaching maturity. Selection for accelerated development would be so intense that we would expect a shortened period of immaturity to be evolved very rapidly, even if it meant seriously compromising other adaptations. By contrast, if the mortality rate dropped to only 0.5 per *month*, a 2 per cent acceleration would increase fitness by only about 10 per cent. That organisms with high juvenile mortality rates have correspondingly rapid development is a significant generalization that I assume would be universally conceded. Selection for developmental rate must be extremely sensitive to variations in juvenile mortality rate.

The probability of reaching maturity is only one aspect of the mortality probability distribution to which an organism can be expected to adapt. When its normal development takes it through a succession of stages of different mortality rates, it can be expected to hurry through the stages of high mortality, and to proceed slowly (by comparison) through those that are less dangerous. If S_i is the probability per unit time of surviving during stage i, which lasts for the time period t_i, the probability of surviving to stage n would be

$$P_n = (S_1^{t_1})(S_2^{t_2}) \ldots (S_{n-1}^{t_{n-1}})$$

The goal of the Darwinian demon will be to maximize P_n, in whatever way it can be accomplished. It can be expected to maximize ecological adaptation in every stage. This is equivalent to maximizing S_i. It can also be expected to minimize t_i, and the extent of this minimization will vary in accordance with S_i. A given amount of acceleration of development in a high-mortality stage will increase fitness more than it would in a low-mortality stage. We should expect development to be most rapid in stages of high mortality.

This expectation is realized in the life cycles of a large number of plants and animals. Note, for instance, the association of rapid development and high mortality rates in the planktonic young of marine invertebrates that have low mortality rates and low rates of development after successful metamorphosis into the sessile stages. Another interesting example is provided by the birds. Their commitment to a volant adulthood places a great burden of preparatory morphogenesis on the young. The machinery of flight, to be at all effective, must be precisely formed and requires complex structures in all regions of the body. The ecological fitness of nestling birds is severely compromised by the involvement of so much of the body in the production of the machinery of flight. Hence the nestlings are ecologically helpless and have high mortality rates. This helps to account for the rapidity of development up to the stage at which flight becomes possible. Having completed the rushed job of achieving minimum effectiveness in flight, the bird then enters a period of extraordinarily slow development. A number of people have marveled at the way birds, having achieved flight, ecological independence, and nearly full size in a few weeks, nevertheless may not achieve sexual maturity for a long time, perhaps not for several years. I believe that the answer lies in the distribution of mortality rates, and the consequent tendency for selection to accelerate high-mortality stages and to allow the low-mortality stages to endure, by comparison, for a long time. A juvenile bird, capable of flight, and having no reproductive duties to perform, enjoys a very low probability of mortality. It is understandable that selection for rapidity of development in this stage would be greatly relaxed. These considerations apply to typical birds, those that are strong fliers as adults but are helpless and highly vulnerable as nestlings. Birds that depart from this typical condition show the expected departures from the typical distribution of morphogenetic rates. Those that nest in especially safe retreats, such as holes in trees or on cliffs or isolated islands, develop slowly in comparison with other birds. Ground-living birds do not enjoy an extremely safe juvenile period and have this period correspondingly curtailed. The flightless ostrich, the largest of birds, matures as rapidly as some birds not one per cent of its mass. The independent reduction of developmental rate in all groups of hole-nesting birds was reviewed by Haartman (1957). Wynne-Edwards (1962) has surveyed these time relations in the life cycles of birds in general. The conclusions he bases on them are in fundamental opposition to those expressed above, and will be discussed mainly on pp. 244-246.

CHAPTER 4

Group Selection

THIS BOOK is a rejoinder to those who have questioned the adequacy of the traditional model of natural selection to explain evolutionary adaptation. The topics considered in the preceding chapters relate mainly to the adequacy of this model in the realms of physiological, ecological, and developmental mechanisms, matters of primary concern to individual organisms. At the individual level the adequacy of the selection of alternative alleles has been challenged to only a limited degree. Many more doubts on the importance of such selection have been voiced in relation to the phenomena of interactions among individuals. Many biologists have implied, and a moderate number have explicitly maintained, that groups of interacting individuals may be adaptively organized in such a way that individual interests are compromised by a functional subordination to group interests.

It is universally conceded by those who have seriously concerned themselves with this problem (e.g., Allee *et al.*, 1949; Haldane, 1932; Lewontin, 1958B, 1962; Slobodkin, 1954; Wynne-Edwards, 1962; Wright, 1945) that such group-related adaptations must be attributed to the natural selection of alternative *groups* of individuals and that the natural selection of alternative alleles within populations will be opposed to this development. I am in entire agreement with the reasoning behind this conclusion. Only by a theory of between-group selection could we achieve a scientific explanation of group-related adaptations. However, I would question one of the premises on which the reasoning is based. Chapters 5 to 8 will be primarily a defense of the thesis that group-related adaptations do not, in fact, exist. A *group* in this discussion should be understood to mean something other than a family and to be composed of individuals that need not be closely related.

The present chapter examines the logical structure of the theory of selection between groups, but first I wish to consider an apparent exception to the rule that the natural selection of individuals cannot produce group-related adaptations. This exception may be found in animals that live in stable social groups and have the intelligence and other mental qualities necessary to form a system of personal friendships and animosities that transcend the limits of family relationship. Human society would be impossible without the ability of each of us to know, individually, a variety of neighbors. We learn that Mr. X is a noble gentleman and that Mr. Y is a scoundrel. A moment of reflection should convince anyone that these relationships may have much to do with evolutionary success. Primitive man lived in a world in which stable interactions of personalities were very much a part of his ecological environment. He had to adjust to this set of ecological factors as well as to any other. If he was socially acceptable, some of his neighbors might bring food to himself and his family when he was temporarily incapacitated by disease or injury. In time of dearth, a stronger neighbor might rob our primitive man of food, but the neighbor would be more likely to rob a detestable primitive Mr. Y and his troublesome family. Conversely, when a poor Mr. X is sick our primitive man will, if he can, provide for him. Mr. X's warm heart will know the emotion of gratitude and, since he recognizes his benefactor and remembers the help provided, will probably reciprocate some day. A number of people, including Darwin (1896, Chap. 5), have recognized the importance of this factor in human evolution. Darwin speaks of it as the "lowly motive" of helping others in the hope of future repayment. I see no reason why a conscious motive need be involved. It is necessary that help provided to others be occasionally reciprocated if it is to be favored by natural selection. It is not necessary that either the giver or the receiver be aware of this.

Simply stated, an individual who maximizes his friendships and minimizes his antagonisms will have an evolutionary advantage, and selection should favor those characters that promote the optimization of personal relationships. I imagine that this evolutionary factor has increased man's capacity for altruism and compassion and has tempered his ethically less acceptable heritage of sexual and predatory aggressiveness. There is theoretically no limit to the extent and complexity of group-related behavior that this factor could produce, and the immediate goal of such behavior would always be the well-being of some other individual, often genetically unrelated. Ultimately, however, this would not be an adaptation for group benefit. It would be developed by the differential survival of individuals and would be designed for the perpetuation of the genes of the individual providing the benefit to another. It would involve only such immediate self-sacrifice for which the probability of later repayment would be sufficient justification. The natural selection of alternative alleles can foster the production of individuals willing to sacrifice their lives for their offspring, but never for mere friends.

The prerequisites for the operation of this evolutionary factor are such as to confine it to a minor fraction of the Earth's biota. Many animals form dominance hierarchies, but these are not sufficient to produce an evolutionary advantage in mutual aid. A consistent interaction pattern between hens in a barnyard is adequately explained without postulating emotional bonds between individuals. One hen reacts to another on the basis of the social releasers that are displayed, and if individual recognition is operative, it merely adjusts the behavior towards another individual according to the immediate results of past interactions. There is no reason to believe that a hen can harbor grudges against or feel friendship toward another hen. Certainly the repayment of favors would be out of the question.

A competition for social goodwill cannot fail to have been a factor in human evolution, and I would expect that it would operate in many of the other primates. Altman (1962) described the formation of semipermanent coalitions between individuals within bands of wild rhesus monkeys and cited similar examples from other primates. Members of such coalitions helped each other in conflicts and indulged in other kinds of mutual aid. Surely an individual that had a better than average ability to form such coalitions would have an evolutionary advantage over its competitors. Perhaps this evolutionary factor might operate in the evolution of porpoises. This seems to be the most likely explanation for the very solicitous behavior that they sometimes show toward each other (Slijper, 1962, pp. 193-197). I would be reluctant, however, to recognize this factor in any group but the mammalia, and I would imagine it to be confined to a minority of this group. For the overwhelming mass of the Earth's biota, friendship and hate are not parts of the ecological environment, and the only way for socially beneficial self-sacrifice to evolve is through the biased survival and extinction of populations, not by selective gene substitution within populations.

To MINIMIZE recurrent semantic difficulties, I will formally distinguish two kinds of natural selection. The natural selection of alternative alleles in a Mendelian population will henceforth be called *genic selection*. The natural selection of more inclusive entities will be called *group selection*, a term introduced by Wynne-Edwards (1962). *Intrademic* and *interdemic*, and other terms with the same prefixes, have been used to make the same distinction. It has been my experience, however, that the repeated use in the same discussion of "inter" and "intra" for specifically contrasted concepts is a certain cause of confusion, unless a reader exerts an inconvenient amount of attention to spelling, or a speaker indulges in highly theatrical pronunciation.

The definitions of other useful terms, and the conceptual relations between the various creative evolutionary factors and the production of adaptation are indicated in Figure 1. Genic selection should be assumed to imply the current conception of natural selection often termed *neo-Darwinian*. An *organic adaptation* would be a mechanism designed to promote

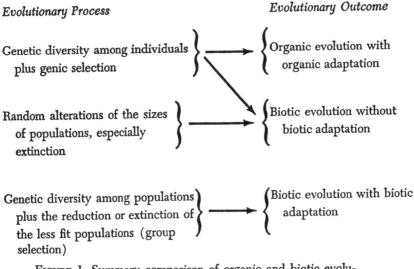

FIGURE 1. Summary comparison of organic and biotic evolution, and of organic and biotic adaptation.

the success of an individual organism, as measured by the extent to which it contributes genes to later generations of the population of which it is a member. It has the individual's *inclusive fitness* (Hamilton, 1964A) as its goal. Biotic evolution is any change in a biota. It can be brought about by an evolutionary change in one or more of the constituent populations, or merely by a change in their relative numbers. A *biotic adaptation* is a mechanism designed to promote the success of a biota, as measured by the lapse of time to extinction. The biota considered would have to be restricted in scope so as to allow comparison with other biotas. It could be a single biome, or community, or taxonomic group, or, most often, a single population. A change in the fish-fauna of a lake would be considered biotic evolution. It could come about through some change in the characters of one or more of the constituent populations or through a change

GROUP SELECTION

in the relative numbers of the populations. Either would result in a changed fish-fauna, and such a change would be biotic evolution. A biotic adaptation could be a mechanism for the survival of such a group as the fish-fauna of a lake, or of any included population, or of a whole species that lives in that lake and elsewhere.

I believe that it is useful to make a formal distinction between biotic and organic evolution, and that certain fallacies can be avoided by keeping the distinction in mind. It should be clear that, in general, the fossil record can be a direct source of information on organic evolution only when changes in single populations can be followed through a continuous sequence of strata. Ordinarily the record tells us only that the biota at time t' was different from that at time t and that it must have changed from one state to the other during the interval. An unfortunate tendency is to forget this and to assume that the biotic change must be ascribed to appropriate organic change. The horse-fauna of the Eocene, for instance, was composed of smaller animals than that of the Pliocene. From this observation, it is tempting to conclude that, at least most of the time and on the average, a larger than mean size was an advantage to an individual horse in its reproductive competition with the rest of its population. So the component populations of the Tertiary horse-fauna are presumed to have been evolving larger size most of the time and on the average. It is conceivable, however, that precisely the opposite is true. It may be that at any given moment during the Tertiary, most of the horse populations were evolving a smaller size. To account for the trend towards larger size it is merely necessary to

GROUP SELECTION

make the additional assumption that group selection favored such a tendency. Thus, while only a minority of the populations may have been evolving a larger size, it could have been this minority that gave rise to most of the populations of a million years later. Figure 2 shows how the same observations on the fossil record can be rationalized on two entirely different bases. The unwarranted assumption of organic evolution as an explanation for biotic evolution dates

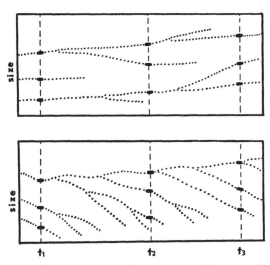

FIGURE 2. Alternative ways of interpreting the same observations of the fossil record. Average sizes in hypothetical horse species at three different times are indicated by boldface marks on the vertical time-scale at times t_1, t_2, and t_3. Upper and lower diagrams show the same observations. In the upper, hypothetical phylogenies explain the observations as the result of the organic evolution of increased size and of occasional chance extinction. In the lower, the hypothetical phylogenies indicate the organic evolution mainly of decreased size, but with effective counteraction by group selection so that the biota evolves a larger average size.

99

at least from Darwin. In *The Origin of Species* he dealt with a problem that he termed "advance in organization." He interpreted the fossil record as indicating that the biota has evolved progressively "higher" forms from the Cambrian to Recent, clearly a change in the biota. His explanation, however, is put largely in terms of the advantage that an individual might have over his neighbors by virtue of a larger brain, greater histological complexity, etc. Darwin's reasoning here is analogous to that of someone who would expect that if the organic evolution of horses proceeded toward larger size during the Tertiary, most equine mutations during this interval must have caused larger size in the affected individuals. I suspect that most biologists would tend toward the opposite view, and expect that random changes in the germ plasm would be more likely to curtail growth than to augment it. Organic evolution would normally run counter to the direction of mutation pressure. There is a formally similar relation between organic evolution and group selection. Organic evolution provides genetically different populations, the raw material on which group selection acts. There is no necessity for supposing that the two forces would normally be in precisely the same direction. It is conceivable that at any given moment since the Cambrian, the majority of organisms were evolving along lines that Darwin would consider retrogression, degeneration, or narrow specialization, and that only a minority were progressing. If the continued survival of populations were sufficiently biased in favor of this minority, however, the biota as a whole might show "progress" from one geologic period to the next. I expect that the fossil record is actually of little use in

GROUP SELECTION

evaluating the relative potency of genic and group selection.

In another respect the analogy between mutation and organic evolution as sources of diversity may be misleading. Mutations occur at random and are usually destructive of any adaptation, whereas organic evolution is largely concerned with the production or at least the maintenance of organic adaptation. Any biota will show a system of adaptations. If there is no group selection, i.e., if extinction is purely by chance, the adaptations shown will be a random sample of those produced by genic selection. If group selection does operate, even weakly, the adaptations shown will be a biased sample of those produced by genic selection. Even with such bias in the kinds of adaptations actually represented, we would still recognize genic selection as the process that actually produced them. We could say that the adaptations were produced by group selection only if it was so strong that it constantly curtailed organic evolution in all but certain favored directions and was thus able, by its own influence, to accumulate the func- of organic adaptations and the production of the ad- tional details of complex adaptations. This distinction between the production of a biota with a certain set aptations of a biota will be emphasized again in a number of contexts.

IN DISCUSSING the adaptations and the general fitness of an individual organism we can often make value judgments with some confidence. We may be familiar enough with the physiology and ecology of the organism to state an opinion on how fit it is, relative to other organisms in the same population. We can be especially confident of our opinion when the organism shows some gross impairment of what we know to be an important mechanism. A horse with a broken leg has a very low fitness compared with most of the uninjured members of its herd. It might happen that soon after such an unfavorable estimate of its fitness, all but two of the herd-mates of our injured horse wander into a steep-walled canyon where they are trapped and killed by a fire. It might turn out that, by chance, the two that escaped are a son and a daughter of the individual with the broken leg. This would make this horse extraordinarily successful but would not invalidate our adverse judgment of its fitness. We could still insist that it was of very low fitness but also very lucky. Fitness is not related to genetic survival in any deterministic fashion. Chance is also an important factor. We cannot measure fitness by evolutionary success on an individual basis. It should be judged in individuals by the extent and effectiveness of design for survival. We can judge any horse with a broken leg to be of quite low fitness, because such an injury is a grave impairment of its adaptive design. Only if such judgments conflict with the facts of evolutionary success and failure in a significant majority of cases would we be proved wrong. Such judgments are undoubtedly right most of the time. For a horse population we can surely say that individuals characterized by fleetness, disease resistance, sensory acuity, and fertility are more fit than those that are less fleet, less resistant, etc. The science of equine physiology has advanced enough so that we have a detailed understanding and appreciation of the horse's design for survival and can have no reasonable doubt of the importance and effectiveness of that design.

GROUP SELECTION

Ideally we should approach the subject of biotic adaptation in the same way. Unfortunately we do not have a comparable understanding of the physiology and ecology of populations and more inclusive groups. Can we describe any biotically adaptive mechanisms of a population of horses? Would we be able to recognize an impairment of any such mechanism as confidently as we would recognize a fracture as an individual impairment? Such questions will be left primarily for later chapters. At this point I wish merely to point out that if it were not for the obvious existence of organic adaptations and of their taxonomic diversity, there would be no need for the theory of genic selection. Similarly, unless there are such things as biotic adaptations, there is no need for the theory of group selection.

If we cannot adequately detect and measure design for success, perhaps we can measure the less interesting but necessarily correlated factor of success itself. It is certainly possible, under the proper conditions, to measure the evolutionary success of an individual organism. We need merely count its descendants of two or three generations later and compare this count with the mean of its contemporaries' descendants. More often than not, a highly successful organism would have been of above average fitness. We could determine what characters contribute most to fitness by noting those that are most strongly correlated with success. For populations this method would be more difficult, because a much longer time would be necessary. There is more hope, perhaps, for meaningful answers to the question, "How well is the population now succeeding?" than for the question, "How well will it have succeeded a thousand years from now?" When we see an adult animal or plant that we judge to be healthy and vigorous, and that we know has produced healthy offspring, we regard it as being currently successful. Might there not be comparable indications of health and vigor in a population on the basis of which we might say that it is currently successful and therefore probably fit? This seems to be the most frequent approach to the problem of assaying the presence of biotic adaptation in Mendelian populations.

A commonly assumed measure of population success and well-being is simply the number of individuals. This parameter may sometimes be satisfying and useful, e.g., in comparisons of genetically different fruit-fly populations in identical environments (Carson, 1961). In its crude form, however, I doubt that many people would find this definition consistently acceptable. Is a fox population less successful than the more numerous rabbits on which it feeds? The use of mere number as an indication of adaptation can be made more generally valid by the introduction of various "adjustments." Mass, rather than number, may be used for species of different sized individuals; or comparisons may be limited to ecologically equivalent types, to equal areas, to similar stages in the life cycle, etc. Even then, however, there will be many instances of comparisons that will rank one population as better adapted than another but in so doing will do violence to an intuitive judgment of population fitness. Are the diatom populations of the North Atlantic better adapted than those of Lake Geneva? They are larger populations at any season and they are denser at most. There is no way of knowing just

GROUP SELECTION

where to stop in compiling a list of adjustments and correction factors.

A possible alternative is to use as a measure of success, not population size, but its current rate of change in size (Fisher's *Malthusian parameter*), as was done by Odum and Allee (1956), Kimura (1958), and Barker (1963). A population that is increasing rapidly would be considered more successful than one that is stationary in size or decreasing. Here again there are situations in which the approach would not agree with what we feel we should really mean by population well-being. By this criterion the diatom populations of both the North Atlantic and Lake Geneva would be alternatively at a much higher and a much lower level of adaptation than the human population. A number of treatments of the problem (e.g., Pimentel, 1961; Brereton, 1962A; Wynne-Edwards, 1962) have assumed, in fact, that there are times when a reduction in number is adaptive. This conclusion is implied by the use of the term *regulation* in the sense of adaptive control of population size, a matter taken up in Chapter 8. Brown (1958) maintained that the best-adapted populations show, not stability, but wide fluctuations in number, which make it possible for the species to expand into new habitats.

Lewontin (1958B) suggested that population fitness be measured by ecological versatility. Thus if one population can survive in only one of two environments and another population can survive in both, he would regard the second as better adapted. Here again the suggestion seems reasonable at first but breaks down in any number of actual situations. If we were to recognize three general habitats, the terrestrial, the freshwater, and the marine, the populations of certain euryhaline or amphibious animals would have to be considered better adapted than the great majority of the birds and mammals. Many bacteria would be better adapted than any angiosperm, and so on. Judgments based on habitat versatility would depend as much on the classification of habitats as on the properties of populations.

The basic problem, which Lewontin clearly recognized, is the very common one of the easily measured variables' not being the theoretically important ones. The important factor is the degree of assurance, especially that provided by biotic adaptations, for long-term population survival. Thoday (1953, 1958) proposed this as the definition of population fitness, but did not suggest any simple formula for its objective measurement. The attainment of such population fitness is often implied as an important element in evolutionary progress (see pp. 49-54). In practice, estimates of current population success (itself a crude estimate of fitness, or design for success) are based on readily measurable demographic variables that must be very imperfectly correlated with long-term survival and extinction. It may be true that the set, "large population," has a lower rate of extinction within 10,000 years than the set, "small population." In the same way increasing populations can be given more favorable prognoses than diminishing populations. I am sure that most biologists would agree, however, that such characteristics form an extremely unreliable basis for predicting long-term survival and extinction.

My own preference is for numerical stability, regardless of absolute numbers, as a measure of current population success. This measure is assumed to be the most relevant by Pimentel (1961), Brereton (1962A and B), Dunbar (1960), and Wynne-Edwards (1962). Whatever advantages this measure possesses may be partly offset by its being somewhat more difficult to assess than either absolute size or momentary rate of change in size. Stability can be assessed by the amplitude of fluctuation about a long-term mean. A population with a very low amplitude of fluctuation would be considered more successful than one that has large ups and downs. It is the downs that are important. A population that does not, over a long interval of time, drop below one-half of its long-term mean would presumably be in a healthier state than one that frequently drops below one-third of its long-term mean. The important consideration is the likelihood of dropping to zero, and I presume that this is more likely for the relatively more variable population, regardless of its absolute numbers. The cause of such stability is supposed by Brereton (1962A) and Wynne-Edwards (1962) to be due in most cases to the presence of biotic adaptations. Some restriction would have to be placed on the way in which fluctuations are measured. Those that are a part of the normal life cycle should not be included. Comparisons should be made between means of entire cycles or of similar points (e.g., minima) of successive cycles.

All of the various criteria mentioned above have been assumed by different authors to be reliable measures of group success and to be the obvious goals of biotic adaptations. So all must be borne in mind in the ensuing chapters where these supposed adaptations are discussed. In this chapter the treatment of biotic adaptation will be confined to the theory of its genesis, and will indicate that there is no firm reason to expect group selection to be an important creative factor.

First of all, it is essential, before proceeding further with the discussion, that the reader firmly grasp the general meaning of biotic adaptation. He must be able to make a conceptual distinction between a population of adapted insects and an adapted population of insects. The fact that an insect population survives through a succession of generations is not evidence for the existence of biotic adaptation. The survival of the population may be merely an incidental consequence of the organic adaptations by which each insect attempts to survive and reproduce itself. The survival of the population depends on these individual efforts. To determine whether this survival is the proper function or merely an incidental by-product of the individual effort must be decided by a critical examination of the reproductive processes. We must decide: Do these processes show an effective design for maximizing the number of descendants of the individual, or do they show an effective design for maximizing the number, rate of growth, or numerical stability of the population or larger system? Any feature of the system that promotes group survival and cannot be explained as an organic adaptation can be called a biotic adaptation. If the population has such adaptations it can be called an adapted population. If it does not, if its continued survival is merely incidental to the operation of organic adaptations, it is merely a population of adapted insects.

GROUP SELECTION

LIKE the theory of genic selection, the theory of group selection is logically a tautology and there can be no sane doubt about the reality of the process. Rational criticism must center on the importance of the process and on its adequacy in explaining the phenomena attributed to it. An important tenet of evolutionary theory is that natural selection can produce significant cumulative change only if selection coefficients are high relative to the rates of change of the selected entity. Since genic selection coefficients are high relative to mutation rates, it is logically possible for the natural selection of alternative alleles to have important cumulative effects. It was pointed out on pp. 22-23 that there can be no effective selection of somata. They have limited life spans and (often) zero biotic potential. The same considerations apply to populations of somata. I also pointed out that genotypes have limited lives and fail to reproduce themselves (they are destroyed by meiosis and recombination), except where clonal reproduction is possible. This is equally true of populations of genotypes. All of the genotypes of fruit-fly populations now living will have ceased to exist in a few weeks. Within a population, only the gene is stable enough to be effectively selected. Likewise in selection among populations, only populations of genes (gene pools) seem to qualify with respect to the necessary stability. Even gene pools will not always qualify. If populations are evolving rapidly and have a low rate of extinction and replacement, the rate of endogenous change might be too great for group selection to have any cumulative effect. This argument precisely parallels that which indicates that mutation rates must be low relative to selection coefficients for genic selection to be effective.

If a group of adequately stable populations is available, group selection can theoretically produce biotic adaptations, for the same reason that genic selection can produce organic adaptations. Consider again the evolution of size among Tertiary horses. Suppose that at one time there was a genus of two species, one that averaged 100 kilograms when full grown and another that averaged 150 kilograms. Assume that genic selection in both species favored a smaller size so that a million years later the larger of the two averaged only 130 kilograms and the smaller had become extinct, but had lost 20 kilograms before it did so. In this case we could say that the genus evolved an increased size, even though both of the included species evolved a decreased size. If the extinction of the smaller species is not just a chance event but is attributable to its smaller size, we might refer to large size as a biotic adaptation of a simple sort. However, it is the origin of complex adaptations, for which the concept of functional design would be applicable, that is the important consideration.

If alternative gene pools are not themselves stable, it is still conceivable that group selection could operate among more or less constant rates of change. A system of relatively stable rates of change in the gene frequencies of a population might be called an evolutionary trajectory. It could be described as a vector in n-dimensional space, with n being the number of relevant gene frequencies. In a given sequence of a few generations a gene pool may be undergoing certain kinds of change at a certain rate. This is only one of an infinite number of other evolutionary trajectories that might conceivably be followed. Some trajectories may be more likely to lead to extinction than others, and group selection will then operate by allowing different kinds of evolutionary change to continue for different average lengths of time. There is paleontological evidence that certain kinds of evolutionary change may continue for appreciable lengths of time on a geological scale. Some of the supposed examples disappear as the evidence accumulates and shows that actual courses of evolution are more complex than they may have seemed at first. Other examples are apparently real and are attributed by Simpson (1944, 1953) to continuous genic selection in certain directions, a process he terms "orthoselection."

Wright (1945) proposed that group selection would be especially effective in a species that was divided up into many small populations that were almost but not quite isolated from each other. Most of the evolutionary change in such a species would be in accordance with genic selection coefficients, but the populations are supposed to be small enough so that genes would occasionally be fixed by drift in spite of adverse selection within a population. Some of the genes so fixed might benefit the population as a whole even though they were of competitive disadvantage within the population. A group so favored would increase in size (regarded as a benefit in Wright's discussion) and send out an augmented number of emigrants to neighboring populations. These migrants would partly or wholly counteract the adverse selection of the gene in neighboring populations and give them repeated opportunity for the chance fixation of the gene. The oft-repeated operation of this process eventually would produce complex adaptations of group benefit, but of competitive disadvantage to an individual. According to this theory, selection not only can act on preexisting variation, but also can help to produce the variation on which it acts, by repeatedly introducing the favored gene into different populations.

Wright formally derived this model in a review of a book by G. G. Simpson. Later, Simpson (1953, pp. 123; 164-165) briefly criticized Wright's theory by pointing out that it leaves too much to a rather improbable concatenation of the population parameters of size, number, degree of isolation, and the balance of genic and group selection coefficients. The populations have to be small enough for genetic drift to be important, but not so small that they are in danger of extinction, and they have to be big enough for certain gene substitutions to be more important than chance factors in determining size and rate of emigration. The unaugmented rates of immigration must be too small to reestablish the biotically undesirable gene after it is lost by drift. The populations must be numerous enough for the postulated process to work at a variety of loci, and each of the populations must be within the necessary size range. Lastly, the balance of these various factors must persist long enough for an appreciable amount of evolutionary change to take place. At the moment, I can see no hope of achieving any reliable estimate of how frequently the necessary conditions have been realized, but surely the frequency of such combinations of circumstances must be relatively low and the combinations quite temporary when they do occur. Simpson also expressed doubts on the reality of the biotic adaptations that Wright's theory was proposed to explain.

GROUP SELECTION

A number of writers have since postulated a role for the selection of alternative populations within a species in the production of various supposed "altruistic" adaptations. Most of these references, however, have completely ignored the problem that Wright took such pains to resolve. They have ignored the problem of how whole populations can acquire the necessary genes in high frequency in the first place. Unless some do and some do not, there is no set of alternatives for group selection to act upon. Wright was certainly aware, as some later workers apparently were not, that even a minute selective disadvantage to a gene in a population of moderate size can cause an almost deterministic reduction of the gene to a negligible frequency. This is why he explicitly limited the application of his model to those species that are subdivided into many small local populations with only occasional migrants between them. Others have postulated such group selection as an evolutionary factor in species that manifestly do not have the requisite population structures. Wynne-Edwards (1962), for example, postulated the origin of biotic adaptations of individual disadvantage, by selection among populations of smelts, in which even a single spawning aggregation may consist of tens of thousands of individuals. He envisioned the same process for marine invertebrates that may exist as breeding adults by the million per square mile and have larval stages that may be dispersed many miles from their points of origin.

A POSSIBLE escape from the necessity of relying on drift in small populations to fix the genes that might contribute to biotic adaptation, is to assume that such genes are not uniformly disadvantageous in competitive individual relationships. If such a gene were, for some reason, individually advantageous in one out of ten populations, group selection could work by making the descendants of that population the sole representatives of the species a million years later. However, this process also loses plausibility on close examination. Low rates of endogenous change relative to selection coefficients are a necessary precondition for any effective selection. The necessary stability is the general rule for genes. While gene pools or evolutionary trajectories can persist little altered through a long period of extinction and replacement of populations, there is no indication that this is the general rule. Hence the effectiveness of group selection is open to question at the axiomatic level for almost any group of organisms. The possibility of effective group selection can be dismissed for any species that consists, as many do, of a single population. Similarly the group selection of alternative species cannot direct the evolution of a monotypic genus, and so on.

Even in groups in which all of the necessary conditions for group selection might be demonstrated, there is no assurance that these conditions will continue to prevail. Just as the evolution of even the simplest organic adaptation requires the operation of selection at many loci for many generations, so also would the production of biotic adaptation require the selective substitution of many groups. This is a major theoretical difficulty. Consider how rapid is the turnover of generations in even the slowest breeding organisms, compared to the rate at which populations replace each other. The genesis of biotic adaptation must for this reason be orders of magnitude slower than that of organic adaptation. Genic selection may take the form of the replacement of one allele by another at the rate of 0.01 per generation, to choose an unusually high figure. Would the same force of group selection mean that a certain population would be 0.01 larger, or be growing 0.01 faster, or be 0.01 less likely to become extinct in a certain number of generations, or have a 0.01 greater emigration rate than another population? No matter which meaning we assign, it is clear that what would be a powerful selective force at the genic level would be trivial at the group level. For group selection to be as strong as genic selection, its selection coefficients would have to be much greater to compensate for the low rate of extinction and replacement of populations.

The rapid turnover of generations is one of the crucial factors that makes genic selection such a powerful force. Another is the large absolute number of individuals in even relatively small populations, and this brings us to another major difficulty in group selection, especially at the species level. A species of a hundred different populations, sufficiently isolated to develop appreciable genetic differences, would be exceptional in most groups of organisms. Such a complexly subdivided group, however, might be in the same position with respect to a bias of 0.01 in the extinction and replacement of groups, as a population of 50 diploid individuals with genic selection coefficients that differ by 0.01. In the population of 50 we would recognize genetic drift, a chance factor, as much more important than selection as an evolutionary force. Numbers of populations in a species, or of taxa in higher categories, are usually so small that chance would be much more important in determining group survival than would even relatively marked genetic differences among the groups. By analogy with the conclusions of population genetics, group selection would be an important creative force only where there were at least some hundreds of populations in the group under consideration.

Obviously the comments above are not intended to be a logically adequate evaluation of group selection. Analogies with the conclusions on genic selection are only analogies, not rigorously reasoned connections. I would suggest, however, that they provide a reasonable basis for skepticism about the effectiveness of this evolutionary force. The opposite tendency is frequently evident. A biologist may note that, logically and empirically, the evolutionary process is capable of producing adaptations of great complexity. He then assumes that these adaptations must include not only the organic but also the biotic, usually discussed in such terms as "for the good of the species." A good example is provided by Montagu (1952), who summarized the modern theory of natural selection and in so doing presented an essentially accurate picture of selective gene substitution by the differential reproductive survival of individuals. Then in the same work he states, "We begin to understand then, that evolution itself is a process which favors cooperating rather than disoperating groups and that 'fitness' is a function of the group as a whole rather than separate individuals." This kind of evolution and fitness is attributed to the previously described natural selection of individuals. Such an extrapolation from conclusions based on analyses of the possibilities of selective gene substitutions *in* populations to the production of biotic adaptations *of* populations is

GROUP SELECTION

entirely unjustified. Lewontin (1961) has pointed out that population genetics as it is known today relates to genetic processes in populations, not of populations.

LEWONTIN (1962; Lewontin and Dunn, 1960) has produced what seems to me to be the only convincing evidence for the operation of group selection. There is a series of alleles symbolized by t in house-mouse populations that produces a marked distortion of the segregation ratio of sperm. As much as 95 per cent of the sperm of a heterozygous male may bear such a gene, and only 5 per cent bear the wild-type allele. This marked selective advantage is opposed by other adverse effects in the homozygotes, either an embryonic lethality or male sterility. Such characters as lethality, sterility, and measurable segregation ratios furnish an excellent opportunity for calculating the effect of selection as a function of gene frequency in hypothetical populations. Such calculations, based on a deterministic model of selection, indicate that these alleles should have certain equilibrium frequencies in the populations in which they occur. Studies of wild populations, however, consistently give frequencies below the calculated values. Lewontin concludes that the deficiency must be ascribed to some force in opposition to genic selection, and that group selection is the likely force. He showed that by substituting a stochastic model of natural selection, so as to allow for a certain rate of fixation of one or another allele in family groups and small local populations, he could account for the observed low frequencies of the t-alleles.

It should be emphasized that this example relates

to genes characterized by lethality or sterility and extremely marked segregation distortions. Selection of such genes is of the maximum possible intensity. Important changes in frequency can occur in a very few generations as a result of genic selection, and no long-term isolation is necessary. Populations so altered would then be subject to unusually intense group selection. A population in which a segregation distorter reaches a high frequency will rapidly become extinct. A small population that has such a gene in low frequency can lose it by drift and thereafter replace those that have died out. Only one locus is involved. One cannot argue from this example that group selection would be effective in producing a complex adaptation involving closely adjusted gene frequencies at a large number of loci. Group selection in this example cannot maintain very low frequencies of the biotically deleterious gene in a population because even a single heterozygous male immigrant can rapidly "poison" the gene pool. The most important question about the selection of these genes is why they should produce such extreme effects. The segregation distortion makes the genes extremely difficult to keep at low frequency by either genic or group selection. Why has there not been an effective selection of modifiers that would reduce this distortion? Why also has there not been effective selection for modifiers that would abolish the lethality and sterility. The t-alleles certainly must constitute an important part of the genetic environment of every other gene in the population. One would certainly expect the other genes to become adapted to their presence.

Segregation distortion is something of a novelty in

natural populations. I would be inclined to attribute the low frequency of such effects to the adjustment of each gene to its genetic environment. When distorter genes appear they would be expected to replace their alleles unless they produced, like the t-alleles, drastic reductions in fitness at some stage of development. When such deleterious effects are mild, the population would probably survive and would gradually incorporate modifiers that would reduce the deleterious effects. In other words, the other genes would adjust to their new genetic environment. It is entirely possible, however, that populations and perhaps entire species could be rendered extinct by the introduction of such genes as the t-alleles of mice. Such an event would illustrate the production, by genic selection, of characters that are highly unfavorable to the survival of the species. The gene in question would produce a high phenotypic fitness in the gamete stage. It might have a low effect on some other stage. The selection coefficient would be determined by the mean of these two effects relative to those of alternative alleles, regardless of the effect on population survival. I wonder if anyone has thought of controlling the mouse population of an area by flooding it with t-carriers.

I AM entirely willing to concede that the kinds of adaptations evolved by a population, for instance segregation distortion, might influence its chance for continued survival. I question only the effectiveness of this extinction-bias in the production and maintenance of any adaptive mechanisms worthy of the name. This is not the same as denying that extinction can be an important factor in biotic evolution. The conclusion is inescapable that extinction has been extremely important in producing the Earth's biota as we know it today. Probably only on the order of a dozen Devonian vertebrates have left any Recent descendants. If it had happened that some of these dozen had not survived, I am sure that the composition of today's biota would be profoundly different.

Another example of the importance of extinction can be taken from human evolution. The modern races and various extinct hominids derive from a lineage that diverged from the other Anthropoidea a million or perhaps several million years ago. There must have been a stage in which man's ancestors were congeneric with, but specifically distinct from, the ancestors of the modern anthropoid apes. At this time there were probably several and perhaps many other species in this genus. All but about four, however, became extinct. One that happened to survive produced the gibbons, another the orang, another the gorilla and chimpanzee, and another produced the hominids. These were only four (or perhaps three or five) of an unknown number of contemporary Pliocene alternatives. Suppose that the number had been one less, with man's ancestor being assigned to the group that became extinct! We have no idea how many narrow escapes from extinction man's lineage may have experienced. There would have been nothing extraordinary about his extinction; on the contrary, this is the statistically most likely development. The extinction of this lineage would, however, have provided the world today with a strikingly different biota. This one ape, which must have had a somewhat greater than average tendency toward bipedal locomotion and, according to recent views, a tendency towards predatory pack behavior, was transferred by evolution from an ordinary animal, with an ordinary existence, to a cultural chain reaction. The production and maintenance of such tributary adaptations as an enlarged brain, manual dexterity, the arched foot, etc. was brought about by the gradual shifting of gene frequencies at each genetic locus in response to changes in the genetic, somatic, and ecological environments. It was this process that fashioned a man from a beast. The fashioning was not accomplished by the survival of one animal type and the extinction of others.

I would concede that such matters of extinction and survival are extremely important in biotic evolution. Of the systems of adaptations produced by organic evolution during any given million years, only a small proportion will still be present several million years later. The surviving lines will be a somewhat biased sample of those actually produced by genic selection, biased in favor of one type of adaptive organization over another, but survival will always be largely a matter of historical accident. It may be that some people would not even recognize such chance extinction as important in biotic evolution. Ecologic determinists might attribute more of a role to the niche factor; man occupies an ecologic niche, and if one ancestral ape had failed to fill it, another would have. This sort of thinking probably has some validity, but surely historical contingency must also be an important factor in evolution. The Earth itself is a unique historical phenomenon, and many unique geological and biological events must have had a profound effect on the nature of the world's biota.

There is another example that should be considered, because it has been used to illustrate a contrary point of view. The extinction of the dinosaurs may have been a necessary precondition to the production of such mammalian types as elephants and bears. This extinction, however, was not the creative force that designed the locomotor and trophic specializations of these mammals. That force can be recognized in genic selection in the mammalian populations. There are analogies in human affairs. In World War II there was a rubber shortage due to the curtailment of imports of natural rubber. Scientists and engineers were thereby stimulated to develop suitable substitutes, and today we have a host of their inventions, some of which are superior to natural rubber for many uses. Necessity may have been the mother of invention, but she was not the inventor. I would liken the curtailment of imports, surely not a creative process, to the extinction of the dinosaurs, and the efforts of the scientists and engineers, which certainly were creative, to the selection of alternative alleles within the mammalian populations. In this attitude I ally myself with Simpson (1944) and against Wright (1945), who argued that the extinction of the dinosaurs, since it may have aided the adaptive radiation of the mammals, should be regarded as a creative process.

GROUP selection is the only conceivable force that could produce biotic adaptation. It was necessary, therefore, in this discussion of biotic adaptation to examine the nature of group selection and to attempt some preliminary evaluation of its power. The issue, however, cannot be resolved on the basis of hypothetical examples and appeals to intuitive judgments

GROUP SELECTION

as to what seems likely or unlikely. A direct assessment of the importance of group selection would have to be based on an accurate knowledge of rates of genetic change, due to different causes, within populations; rates of proliferation and extinction of populations and larger groups; relative and absolute rates of migration and interbreeding; relative and absolute values of the coefficients of genic and group selection; etc. We would need such information for a large and unbiased sample of present and past taxa. Obviously this ideal will not be met, and some indirect method of evaluation will be necessary. The only method that I can conceive of as being reliable is an examination of the adaptations of animals and plants to determine the nature of the goals for which they are designed. The details of the strategy being employed will furnish indications of the purpose of its employment. I can conceive of only two ultimate purposes as being indicated, genic survival and group survival. All other kinds of survival, such as that of individual somata, will be of the nature of tactics employed in the grand strategy, and such tactics will be employed only when they do, in fact, contribute to the realization of a more general goal.

The basic issue then is whether organisms, by and large, are using strategies for genic survival alone, or for both genic and group survival. If both, then which seems to be the predominant consideration? If there are many adaptations of obvious group benefit which cannot be explained on the basis of genic selection, it must be conceded that group selection has been operative and important. If there are no such adaptations, we must conclude that group selection has not been important, and that only genic selection—natural selection in its most austere form—need be recognized as the creative force in evolution. We must always bear in mind that group selection and biotic adaptation are more onerous principles than genic selection and organic adaptation. They should only be invoked when the simpler explanation is clearly inadequate. Our search must be specifically directed at finding adaptations that promote group survival but are clearly neutral or detrimental to individual reproductive survival in within-group competition. The criteria for the recognition of these biotic adaptations are essentially the same as those for organic adaptations. The system in question should produce group benefit in an economical and efficient way and involve enough potentially independent elements that mere chance will not suffice as an explanation for the beneficial effect.

Chapters 5 to 8 are a review of what are apparently regarded as the more likely examples of biotic adaptation. I will discuss these various examples in an attempt to evaluate their reality and thereby assess the importance of group selection as a creative evolutionary force that supplements genic selection.

CHAPTER 5

Adaptations of the Genetic System

THE MACHINERY of sexual reproduction in higher animals and plants is unmistakably an evolved adaptation. It is complex, remarkably uniform, and clearly directed at the goal of producing, with the genes of two parental individuals, offspring of diverse genotypes. How the production of diverse rather than uniform offspring contributes to the ultimate goal of reproductive survival may not be immediately obvious, but the precision of the machinery can only be explained on the basis of selection for efficiency in the production of offspring with the parental genes but not the parental genotypes.

There are some troublesome terminological problems confronting anyone discussing sexual reproduction. The definition used here is implied above. Reproduction is sexual if it produces offspring with new combinations of the parental genes, and does so by means of machinery designed to produce that result. Mutations in asexual clones may ultimately produce genetically diverse descendants, but mutations, as will be argued below, are never by design. So mutation is not a mechanism of sexual reproduction. In the ensuing discussion, a Mendelian population is to be thought of as a group of organisms which, by virtue of at least occasional sexuality, possesses a common gene pool. A narrower category would be a strictly sexual population which, like man, reproduces only by sexual means.

ADAPTATIONS OF THE GENETIC SYSTEM

The production, by a diploid organism, of genetically diverse haploid gametes, which more or less immediately combine with other gametes to form zygotes, can be thought of as a single process of sexual reproduction. According to the definition in the last paragraph, however, the sequence as described includes two sexual mechanisms. Both meiosis and syngamy produce nuclear compositions different from those of the immediately preceding stages. Other organisms, the ferns for example, have the two sexual processes in different parts of the life cycle. The haploid product of meiosis is not physiologically a gamete. It is a dormant spore, a dissemination stage, that can germinate somewhere remote from the parent and develop into an "adult" gametophyte quite different from the sporophyte that produced it. The gametophyte then produces what are functionally gametes that combine with others to form the zygotes that can develop into the ferns a florist would recognize. Here I would be inclined to recognize both the production of spores and the production of zygotes as sexual processes, because both produce individuals genetically different from their parents. Fern sporophytes also reproduce asexually, by the growth of new individuals, genetically identical to the parents, from underground stolons.

The definition, however, excludes diploid parthenogenesis from the category of sexual reproduction. Boyden (1953; 1954) argued that, unlike budding or fission, parthenogenesis proceeds from what are clearly gametes (*sex* cells), by homology with other gametes in other sexual organisms, and that parthenogenesis must therefore be regarded as sexual to distinguish it from budding. I fail to see why sex must be defined on the basis of the structural elements involved. To insist that a process must be called sexual because it makes use of egg cells is comparable to insisting that anything done with the feet must be called locomotion, or that anything done with the mouth must be called eating. Sexual reproduction, like locomotion or feeding, is an effect produced, not a matter of structural homology. There is a need to distinguish, by coining appropriate terms if necessary, between the primitive asexual reproduction of hydra, and the secondary asexuality of daphnid parthenogenesis. To do this by saying that diploid parthenogenesis is really sexual, however, would be like saying that adult barnacles are really motile, because they have and make use of what are obviously legs. These structures are legs by homology with those of other crustaceans, and it is important to distinguish between the primitively sessile condition in, for instance, a sponge and the derived condition of the adult barnacle. The distinction is readily made by using such adjectives as *secondary* or *derived* for the sessile state of the barnacle, and I assume that parthenogenesis can be distinguished from budding in the same way.

BIOLOGISTS very commonly regard sexual reproduction as a biotic adaptation. This is implied in the frequent recognition of sexuality as having the function of providing evolutionary plasticity. The variety of genotypes produced by sexual reproduction presumably provides the species with at least some individuals that can successfully survive any change that is likely to take place. It is difficult to evaluate the extremity of some of the points of view that have been expressed. It sometimes seems as if natural selection itself were regarded as an adaptive mechanism whereby species avoid extinction or exploit the possibilities of adaptive radiation. If provision for geologically future contingencies is considered an evolved adaptation, certainly some sort of teleological thinking is suggested. Such thinking is not necessarily implied by an author who expresses the belief that sexuality decreases the likelihood of extinction. Only if he believes that the need to avoid extinction is a cause of the development or retention of sexuality would his thinking probably be teleological. Even here, he might avoid this stigma by some reference to group selection as a force for the development of evolutionary plasticity. Occasionally, however, teleological reasoning is unmistakable. Darlington (1958), for instance, speaks of the evolutionary factor of *anticipation* and uses the origin of meiosis and of sexual reproduction as examples. He concludes that the evolutionary force of natural selection acting on the genetic system is (p. 239) "endowed with an unparalleled gift, an automatic property of foresight." He bases his argument of the assumption (pp. 214-215) that "it is impossible to imagine [sexual reproduction] as the result of a gradual accumulation of changes each one of which had value as an adaptation." This is a surprising statement, because Boyden (1953) and Dougherty (1955) had already published carefully reasoned theories based on precisely this "impossible" assumption. Indeed, Boyden argued that meiosis and syngamy were evolved gradually (with each stage being useful to its possessor), not once, but several times independently in unrelated lines of descent.

ADAPTATIONS OF THE GENETIC SYSTEM

Huxley (1958, p. 438) concurred in Darlington's recognition of an anticipatory factor in evolution. He maintained that outcrossing in sexual reproduction represents an immediate sacrifice in precision of adaptation, but is maintained because the resulting variability allows the population to meet possible future needs. According to Huxley, adaptation must be a dual concept, and must include mechanisms for long-term population survival as well as immediate success.

I believe that Boyden's and Dougherty's work illustrates a healthy recent tendency to search for evolutionary forces in immediate circumstances rather than in future needs. The belief that genetic recombination is mainly related to future needs, is, however, still prevalent. The best recent work of this sort is that of Stebbins (1960), who attempted a general synthesis based on the assumption that phylogenetic variation in sexual processes in higher plants can be related to phylogenetic variation in the need for evolutionary plasticity. I am not entirely clear from Stebbins' discussion just how long a term is involved in this capacity of genetic plasticity, but it does seem that he has biotic adaptation in mind. I would find his reasoning entirely acceptable if it were applied to individuals rather than to populations. The reasons that he adduces for certain populations' requiring genetic plasticity more than others apply equally well to the individuals in these populations. For instance, the associations that he recognizes between rates of recombination and such factors as environmental stability and rate of turnover of generations are equally significant at both levels. If successive generations are likely to grow up in very similar environments, a precise adaptation to the prevailing conditions, once achieved, had better be maintained. Such adaptation may be largely dependent on an unstable system of heterosis at many loci. Under these circumstances, a suppression of recombination will be favorable to population survival. However, when the conditions faced by one generation are a poor indication of what the next will face, the best strategy will be to produce a new generation with diverse kinds of competence, so that at least some will be adapted to whatever conditions are actually met. In this case the best thing for the population is to maximize recombination so as to produce a wide variety of genotypes. The same arguments are valid for an individual plant. The individual, having succeeded in attaining maturity in a particular environment, is probably of above average genotypic fitness for that environment. If it is to produce offspring that are likely to grow up under the same conditions, it would do well to make them as much like itself as possible. On the other hand, if its own conditions of life are unreliable as an indication of the conditions its offspring will face, it would be better to utilize recombination to broaden the total competence of the group of offspring produced.

Whenever such arguments work as well at one level as at another, the acceptance of one level and rejection of the other might seem a matter of taste and of little consequence. It is in precisely such a situation, however, that Occam's razor must be used. The principle of adaptation must be recognized at no higher a level of organization than is absolutely necessary. If a phenomenon can be explained as an organic adaptation, it is not permissible to explain it as a biotic adaptation.

This particular issue need not be settled by an argument from parsimony, because there are important phenomena in the reproduction of higher plants, and of animals with similar life cycles, that are inexplicable as biotic adaptations for evolutionary plasticity but are easily understood as organic adaptations. These phenomena are seen when we turn our attention to the distribution of sexual and asexual reproductive processes in a life cycle. From the standpoint of long-term population plasticity all that matters is the amount of genetic recombination per unit time or per generation. It makes no difference where in a reproductive cycle the recombination takes place. It would make no difference, for instance, whether the appearance of sexual aphids came in the spring, summer, or fall, or in more than one season. From the standpoint of individual reproductive survival, however, it makes a great deal of difference. The sexual phase should precede that generation for which the probability of a change of conditions is greatest. Thus we see, among higher plants, the prevalence of asexual reproductive processes to produce young in the immediate vicinity of the parent, and sexual processes to produce the vagile pollen grains and seeds. Ferns produce vegetative copies of themselves in the immediately adjacent soil, but genetically diverse spores as dissemination stages for the production of plants that will inevitably have a way of life quite different from the parent sporophyte. Genetically diverse zygotes are later produced to give rise to a group of offspring that will have quite different ecologies from the parent gametophytes. The same conclusions seem

ADAPTATIONS OF THE GENETIC SYSTEM

to hold among animals with formally similar life cycles (Suomalainen, 1953). Populations of aphids, daphnids, and many other invertebrates continue as asexual clones as long as the young develop immediately in the same habitat as the parent, but the young that will be dispersed through great distances or over long periods of time are sexually produced. Internal parasites of animals reproduce as asexual clones within a given host, but produce genetically diverse zygotes for dissemination to other hosts. These phenomena are readily understood on the basis that the reproduction of each individual is designed to maximize the number of its successful offspring. They are inexplicable as biotic adaptations for ensuring evolutionary plasticity. As biotic adaptations they would be interpretable as having the function of maximizing the size of the next generation, but this interpretation could be rejected on the basis of parsimony. Sex may act to increase population size and it may act to confer evolutionary plasticity, but these are effects, not functions.

I believe that the reliability of the ecological conditions of the parental generation as a basis for predicting the conditions to be faced by the offspring will prove to be generally useful in explaining the distribution of sexual and asexual phases of life cycles and in explaining the incidence of various kinds of restrictions on recombination in sexual processes. Stebbins believes that a decrease in chromosome number and consequent increase in linkage is a mechanism for reducing recombination and should be associated with stable habitats. If this is true we might expect that fishes of stable tropical lowlands would have lower chromosome numbers than fishes of vari-

ADAPTATIONS OF THE GENETIC SYSTEM

able high-latitude or high-altitude habitats. No such tendency is apparent from Makino's (1951) list of chromosome numbers. Chromosome numbers are a readily quantified aspect of a fundamentally important adaptive mechanism of almost all organisms. It is remarkable how little we understand of their general significance. Our understanding of the presumably related phenomenon of linkage is likewise meagre, despite the usual textbook assertion that linkage functions to stabilize favorable genetic combinations. This explanation overlooks the other side of the coin: that if linkage keeps favorable combinations together, it also, to the same extent, keeps unfavorable ones together. Only when crossover rates are very low, relative to the intensity of selection, will linkage result in significant departures from the genotype frequencies that would result from independent assortment. Only one example of such an effect is known for certain (Levitan, 1961). Crossover rates are themselves subject to modification by natural selection (Kimura, 1956), but nothing is really understood of their adaptive significance.

THE EXISTENCE of genetic recombination among the bacteria and viruses, and among all of the major groups of higher organisms, indicates that the molecular basis of sexuality is an ancient evolutionary development. Our understanding of the structure of the DNA molecule makes recombination at this level easy to visualize. In a sense sex is at least as ancient as DNA, but it is most unlikely that DNA dates from the earliest stages of organic evolution. The DNA molecule has all the appearances of an evolved adaptation. It is a highly precise and effective mechanism for the coding of information. This precision and effectiveness must be attributed to a long period of selection for perfection in information transfer. A gene is chemically a very specialized entity. The only kind of change it can tolerate, and still function as a gene, is a very limited number of substitutions among rigidly specified chemical groups. With any other kind of change it would cease to be a gene. This degree of specialization is not found among enzymes. A large variety of substitutions of amino acids alter the quantitative aspects of enzymatic activity without destroying that activity.

The traditional view, expressed in most biology textbooks, is that life originated once, and has ever since been evolving in the well-known neo-Darwinian manner. Sometimes the origin of life is envisioned as the accidental synthesis of a gene in the organic soup, with subsequent mutation and the rapid establishment of genetically isolated lines of descent, each with its unique system of evolved adaptations. This view was perhaps admissible when the gene was thought of as a protein. The extreme specialization of the DNA molecule, however, forces us to regard the nuclear gene as a later development. The early establishment of separate phylogenies is also most unlikely. The independent evolution of identical adaptations must happen very seldom. On this view the (nearly) universal occurrence of fundamental biochemical machinery, such as DNA, ATP, and various enzyme systems, raises a dilemma. Either they were brought to perfection immediately by the original species of organism, or they were independently evolved in all the separate phyletic lines. Neither view is acceptable to serious students of biopoesis

ADAPTATIONS OF THE GENETIC SYSTEM

(origin and early development of life). Effective mechanisms of heredity, and of the structural and immunological mechanisms on which the maintenance of individuality depends, must have required a long period of evolution. Before these mechanisms appeared there could have been no genetically isolated evolutionary lines nor even physically definable individuals. As Ehrensvärd (1962) expressed it, "Life is older than organisms." For a long time life was neither monophyletic nor polyphyletic, it was nonphyletic. Ehrensvärd envisioned a stage at which every body of water could be thought of as a single diffuse organism. Adjacent organisms could fuse as readily as adjacent bodies of water. At this stage there was no selection of alternative individuals or lines of descent. Selection acted only at the level of alternative autocatalytic processes, and even now this is still the fundamental nature of natural selection. From the beginning life must have required the use of available chemical energy in the construction of organic catalysts that were effective in the release of the chemical energy. If a molecule increased the efficiency of this energy conversion it would increase the rate of organic activity, including its own synthesis, in its particular part of Ehrensvärd's living puddle. The earliest mechanisms of heredity must have consisted merely of molecules which had a slight tendency to aid in the synthesis of other, metabolically effective, molecules, and which were sufficiently stable to survive the prevailing changes in the ecological environment, perhaps the occasional drying of the puddles. These protogenes would promote the rapid recovery of vital functions where such processes were subject to periodic interruption. Rapidity of their own synthesis during the metabolically active phases would be favorably selected, and the most effective protogenes would gradually increase in number and in geographic distribution. In this way, evolutionary progress made anywhere in the biosphere would become a property of the entire biota. As Ehrensvärd implied, organic evolution in this phase was an almost deterministic process.

Selection eventually produced genetic systems that were so effective in promoting the homeostasis of surrounding somata, and so specialized in their manner of producing this effect, that the fusion of somewhat different systems would produce adaptively inferior combinations. The maintenance of individuality was then selected for, and life became fragmented into physiologically isolated individuals and genetically separate lines of descent. Nonphyletic life became polyphyletic, but only after the natural selection of competing processes had irrevocably fixed on polypeptides as the metabolic tools, and on polynucleotides as the archive of ecologically effective information. When evolution became polyphyletic it lost its deterministic aspects and became exceedingly stochastic.

The genes must have originally been partly independent entities, resembling some of the plasmagenes and viruses of modern organisms. Undoubtedly each species of gene survived best in a particular set of somatic and ecological conditions, but there were no precise transmission mechanisms by which it could be restricted to the optimum conditions, and as a result a highly specialized gene would be at a disadvantage. Successful genes would necessarily be able to survive and multiply under a wide range of con-

ADAPTATIONS OF THE GENETIC SYSTEM

ditions. Dougherty (1955) states, "We can conceive of evolution as occurring at this stage not as descent from a single line of cells, but as descent through populations of diverse cells in which the transfer of hereditary material between cellular compartments went on continually in an intricate way." Perhaps we can think of viral transduction between bacteria as an illustration of this evolutionary stage.

Specialization would be favored, however, in at least some ecological conditions. It might happen that gene *a* and gene *c* function slightly better in association with gene *b* but not with each other. This would favor the association *a-b-c*, by the positive selection of the variants of *a* and *c* that are most likely to form such an association. This change in the genetic environment of *b* would favor the selection of *b*-alleles that function best in association with *a* and *c*. As long as these associations are beneficial, evolution would proceed towards making them more permanent. If, however, as would be expected in some evolutionary lines, the associations are favorable in some of the normal ecological environments but not in all, the primitive organisms could be expected to adapt to such circumstances. Ecological changes could trigger adaptive responses that would stabilize the *a-b-c* association when it is adaptive, and break them up when it is not. Periods of stable gene associations would alternate with times of dissociation and recombination. We would have the beginnings of a cycle of vegetative growth and sexual reproduction, and right from the start, the observed association of sexuality with environmental change would be established.

As mechanisms for stabilizing such a cycle and of

adjusting it to environmental conditions become perfected, the genetic environment of each gene would become more stable, and genic specialization could then proceed. Association with a and c would become an increasingly reliable aspect of the environment of b and this gene could become more and more specialized for efficiency in that genetic niche. This specialization, in turn, would favor still more precise mechanisms of stability and recombination. Detailed discussions of the probable steps in the development of mitosis, meiosis and diploidy are provided by Dougherty (1955) and Boyden (1953).

I would agree, therefore, with Dougherty s opinion that sexual reproduction is as old as life, in that the most primitive living systems were capable of fusion and of combination and recombination of their autocatalytic particles. Modern organisms have evolved elaborate mechanisms for regulating this primitive power of recombination and for maximizing the benefits to be derived from it. This somatic machinery was gradually perfected by the natural selection of genes on the basis of their ability to reproduce themselves effectively. This effectiveness largely depended on their influences on the somatic machinery on which their reproduction came to depend.

ANOTHER aspect of the reproductive fitness of genes is their stability. Their fitness can only be measured by success in replication. If the individual that bears them dies before reproducing, the genes have failed to reproduce. The production of an a by an A is just as much a failure for A, or more so. It directly adds to the frequency of a competing allele. The fittest possible degree of stability is absolute stability. In

ADAPTATIONS OF THE GENETIC SYSTEM

other words, natural selection of mutation rates has only one possible direction, that of reducing the frequency of mutation to zero.

That mutations should continue to occur after billions of years of adverse selection requires no special explanation. It is merely a reflection of the unquestionable principle that natural selection can often produce mechanisms of extreme precision, but never of perfection. It is not so much that mutations occur that requires the attention of biologists, but that they occur so rarely. Where else in nature can we find complex, biochemically active molecules that can persist, unchanged, for centuries or millennia in aqueous systems at terrestrial temperatures? Not only do they persist, they duplicate themselves with great precision. Surely these features can be interpreted only by postulating a long period of relentless selection for stability.

One frequently hears that natural selection will not produce too low a mutation rate because that would reduce the evolutionary plasticity of the species. It has been argued above that evolutionary plasticity is not an adaptation that can be produced by natural selection. Evolution has probably reduced mutation rates to far below species optima, as the result of unrelenting selection for zero mutation rate in every population. Mutation is, of course, a necessary precondition to continued evolutionary change. So evolution takes place, not so much because of natural selection, but to a large degree in spite of it.

The phenomena associated with genetic influences on mutation rate are important evidence. There are known instances of "mutator genes" that can be attributed to specific loci and have the effect of greatly increasing mutation rates at other loci. These genes are always detected and measured by comparison with "normal" mutation rates, and it would appear that genes that permit high mutation rates in a species are rare in natural populations. This rarity argues for their adverse selection under normal conditions. It also indicates that the rest of the genotype is adjusted to the normal alleles at the mutator loci. Any change in the genetic environment of a gene, for instance by introgression into another species, or even by a shift from outbreeding to inbreeding or vice versa, can be expected to increase the mutation rate of a gene (Darlington, 1958; McClintock, 1951). That mutation rate is at its minimum in the normal genetic environment, can only be explained on the assumption of selection for this minimization in that environment.

Even though selection for mutation rate must always have the same direction, it need not always have the same intensity or effectiveness. In a rapidly evolving population, the effect of selection will be hampered by environmental changes, especially by changes in the genetic environment. In a haploid population a deleterious mutation is not masked by dominant alleles. Adverse selection will be continuous, wherever the gene is present, instead of only in rare homozygotes. So selection for low mutation rates should be more intense in haploid organisms. The available evidence seems to bear out this expectation. Catcheside (1951) lists mutation rates for haploid organisms in the range of 10^{-7} to 10^{-10}. The rates listed in genetics texts for diploid man and fruit flies are 10^{-4} to 10^{-6}. Conclusive evidence, however, would depend on well-reasoned adjustments of time-scales and of the magnitudes of the genetic changes.

In any organism, haploid or diploid, the more mutations there are per generation, the more they will reduce individual fitness. If elephants had the same mutation rates, per unit time, as fruit flies, an elephant population would pick up a great burden of new mutations every generation. Mutation would be such a frequent cause of reduced fitness that there would be intense selection for a lowered mutation rate. This explains why mutation rates in such a wide variety of organisms tend to be much the same per generation. In other words, slow breeding organisms have lower mutation rates per unit of absolute time than organisms with a rapid turnover of generations.

Supporters of the contrary view, that mutation is a biotic adaptation, or at least that mutation rates are not minimized by selection but represent a compromise between the need for stability and the need for ensuring evolutionary potential, are legion (Auerbach, 1956; Buzzati-Traverso, 1954; Ives, 1950; Darlington, 1958; Kimura, 1960; and many others). Ives (1950) specifically maintained that the normal function of genes at the mutator loci is to produce mutations.

THERE is frequent allusion in the evolutionary literature, especially in the elementary treatments of the subject, to dominance as an adaptation for evolutionary plasticity. It permits the genes of diploid organisms to be carried in an unexpressed state as a reserve to be utilized if an environmental change demands a rapid evolutionary response by the population. However, both of the really serious theories that have gained acceptance among geneticists, that of Fisher and that of Wright and Haldane, postulate that dominance is evolved by the selection of alternative alleles and without reference to effects on evolutionary plasticity. These theories and the evidence that bears on them were ably reviewed by Sheppard (1958, Chap. 5).

Fisher's theory proposes that intermediacy will be the normal phenotypic effect of heterozygosis, until the effect is modified by selection. If homozygotes for alternative alleles differ greatly in fitness, there would be favorable selection at other loci for any genes that would modify the heterozygote phenotype in the direction of the more favored homozygote. Dominance, then, would be due to the accumulation of modifiers at many loci. Wright and Haldane suggested the alternative view—a one-locus model— that the dominance of wild-type genes over their deleterious alleles is due to selection favoring those genes that are sufficiently potent to produce a normal phenotype in heterozygous combination with their most frequent alleles.

The fact that the dominance of normal genes over their abnormal alleles is a phenomenon found within populations but not in crosses between species or strongly divergent subspecies (evidence reviewed by Dobzhansky, 1951, pp. 104-105) indicates that the action of "modifiers" is important. This is not really an argument for Fisher's theory because there is no evidence in such observations that the modifiers accumulated after the gene in question became normal for the population. It is, however, an indication that the genetic environment is a crucial factor in the production of dominance, as it is in other types of evolutionary change. Initially, one of the factors in

ADAPTATIONS OF THE GENETIC SYSTEM

the selection of a new allele will be how reliably it produces a favorable phenotypic response in the prevailing genetic environment, which would include prevalent alternative alleles and previously established modifiers at other loci. Then if the new allele becomes the normal one at its locus it will be a part of the genetic environment in which other genes, including modifiers at other loci, will be judged. Its presence may tip the balance of selection at several other loci. It would cause the accumulation of additional modifiers that would shift the heterozygous phenotype further in the direction of the more favorable homozygote, and perhaps even beyond. Kimura (1960) has formally analyzed the problem of specifying the optimum degree of dominance under various conditions. It seems inevitable that both Fisher's process and that envisioned by Wright and Haldane would contribute to this optimization of dominance. A recent argument based on embryological considerations (Crosby, 1963) strongly favors Wright and Haldane's view.

Regardless of which phase one wishes to emphasize, neither process involves anything beyond the selection of alternative alleles in the prevailing genetic, somatic, and ecological environments. Both processes result in an increase in the fitness of individuals. There has been no serious attempt to formulate a theory of dominance as a biotic adaptation to achieve evolutionary plasticity. Such thinking is in evidence only at the level of general summaries of evolutionary principles.

It may, in fact, be questioned whether evolutionary plasticity is even an incidental effect of dominance among the heterozygotes of a diploid population.

ADAPTATIONS OF THE GENETIC SYSTEM

Lerner (1953) has argued that rare recessives have but little evolutionary significance. He believes that the "store" of variability that becomes important when environments change is that carried at loci with more than one allele of appreciable frequency and with heterozygotes exhibiting detectable heterosis. Such frequencies are maintained by the heterozygote advantages, not by mutation pressure. Lerner maintained that the effect of such a genetic structure is to cause a resistance to rapid evolutionary change, an effect quite the opposite to what is usually attributed to heterozygosity in elementary texts. There may be some ecological situations in which division of a species into a number of ecologically differentiated populations, each largely homozygous for different alleles, would be better insurance against extinction than would extensive heterozygosity (Lewontin, 1961). There is no support for a general conclusion that heterozygosity has value as a biotic adaptation, either by allowing rapid evolutionary responses to changing conditions, or by increasing tolerance to long-term changes, or in any other way.

ANOTHER supposed biotic adaptation of the genetic system is introgressive hybridization. Anderson (1953) proposed that natural hybridization, even when very rare and when the resulting offspring are usually inviable or sterile, is an important cause of evolutionary change. Back crosses between the hybrids and the parent types could introduce new genes into the parent populations. The rate of such introgression could be much greater than the rate at which mutation introduces new genes and still be so rare as to be difficult to detect in natural populations. I have

ADAPTATIONS OF THE GENETIC SYSTEM

no reason to doubt that introgressive hybridization may be an important evolutionary factor in some groups. Anderson, however, went further, and referred (1953, p. 290) to the selective advantage of introgression, and frequently implied that the enhancement of evolutionary potential is not merely an effect of introgression, but its normal function. To me this interpretation is inconceivable. The whole reproductive machinery of the higher plants (Anderson's main concern) is obviously directed toward the successful crossing of members of the same species. To postulate that this machinery is intended to fail a part of the time is completely gratuitous. Any machinery fails occasionally. Anderson's approach involves the principle of biotic adaptation where there is need only to postulate the mechanical and statistical results of the occasional failure of organic adaptation. In these mechanisms we can clearly appreciate the role of natural selection at the individual level, which has been explained a number of times, recently by Ehrman (1963). As long as plants that cross with others of a different species leave fewer successful offspring than those that cross only with their own kind, whether the difference results from the partial sterility of the hybrids or from some other factor, there will be selection against the tendency to hybridize. Such selection will produce and preserve the machinery that normally ensures conspecific crossing. Spieth (1958) has shown that the degree of development of behavioral mechanisms for avoiding hybridization is a function of the incidence of related sympatric species, with which hybridization might occur unless there were special mechanisms for its avoidance. Such mechanisms, like other mechanisms, must be subject to occasional malfunction. That the malfunctions may produce long-term benefits does not constitute evidence of adaptation. The recognition of adaptation here involves the mistake of assuming that a beneficial effect is necessarily a function. Evolutionary benefits that result from introgression are produced in spite of natural selection, not because of it.

It has long been known that genetic drift in small populations and the low probability of beneficial mutations may restrict the ability of such populations to respond adaptively to changed conditions. Wilson (1963) recently proposed that persistent smallness is therefore a sort of stress to which a population should adapt by mechanisms designed to counteract the genetic effects of smallness. He searched for such effects in species of ants that have effective population sizes of a few hundred individuals. One type of adaptation that he anticipated was provision for outbreeding. Another was an augmentation of the number of reproductive individuals per colony to something beyond the traditional single fertilized queen. He found that multiple queens were common, but he failed to find provisions for outbreeding. On the contrary, he found that brother-sister matings were frequent. Wilson conceded that population adaptation to persistent smallness was not well demonstrated by his survey of the evidence.

THE last aspect of the genetic system that I will discuss is that of the sex ratio and the sex-determining mechanisms. There is the initial question of why there should be separate sexes in some groups and not in others. This problem could most readily be

ADAPTATIONS OF THE GENETIC SYSTEM

attacked by studies of those groups in which some species are monoecious and others dioecious. Many of the genera and families of higher plants would be suitable subjects for such an investigation. Association between monoecy or dioecy and various life-history features might provide clues to the role of natural selection in determining these conditions. Perhaps certain kinds of ecological and demographic circumstances are more effectively handled by separate sexes than by hermaphrodites, or the issue may depend on physiological relations between parent and seed. For the present discussion I will simply assume that we are dealing with regularly dioecious organisms.

Given this circumstance, one might next inquire as to what would constitute an optimum sex ratio for population survival. I will first assume a strictly sexual population and the common circumstance of one male being able to fertilize the eggs of a large number of females. If we also make the assumption that population survival is favored by making the next generation as large as possible, a fit population would be one with a large female majority. This might not be true if population densities were so low that the finding of mates was a serious problem. At low population densities it would be advantageous to maximize the probability that chance encounters would be between individuals of opposite sex. This would be achieved by a 1.00 sex ratio (by custom, males/females). With these assumptions, we would expect sex ratio as a biotic adaptation to show the pattern indicated in the curve marked "maximize next generation" in Figure 3.

ADAPTATIONS OF THE GENETIC SYSTEM

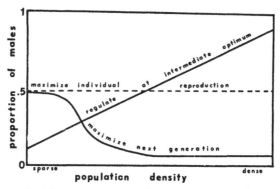

FIGURE 3. Adaptive response of sex ratio to population density. The solid lines are the expected responses on the assumption that sex ratio is a biotic adaptation but with different immediate goals for the adaptation. The dashed line indicates what is expected of sex ratio as the statistical consequence of organic adaptation.

Some discussions of population fitness assume an intermediate optimum size for the next generation. If the population were adapted to achieving this goal we would expect that the fecundity per individual would be high at low population densities, and low at high population densities. With the same assumptions of obligate sexuality and males of multiple potency, we would expect the pattern indicated by the curve marked "regulate at intermediate optimum" in the diagram. At high population densities mostly males would be produced. The abundance of the succeeding generation would be limited to what could be produced by the few females, and this generation would be reduced in size. At low population densities mostly females would be produced. With the postulated high male competence the aggregate fecundity of the mainly female generation would produce a

ADAPTATIONS OF THE GENETIC SYSTEM

greatly increased next generation so as to return the population to its optimum size. These conclusions would differ with different sets of initial assumptions about the population life history and are not important in themselves. They merely indicate various ways in which the sex ratio might be responsive to the needs of populations, and the sorts of things that might be expected on the assumption that sex ratio is a biotic adaptation.

But suppose that the sex ratio is not determined by any consideration of population fitness, but merely by genic selection. The results of such selection would depend on a number of modifying circumstances, but the main consideration would be that originally recognized by R. A. Fisher (1930). In a strictly sexual population, the minority sex has an advantage in passing its genes to the next generation. Every individual has one father and one mother, and the two sexes of the preceding generation must have contributed equally to the present generation. If there was only one male and a hundred females in the preceding generation, that one male must have enjoyed a hundred times the reproductive success of the average female, and in the present generation his genes are a hundred times as numerous as those of the female mean. If this 100:1 female majority is the rule in every generation there would be a selective advantage to any gene that increased the likelihood of its bearer's being a male, or increased the male proportion of the offspring produced. The result would be that the males would become more numerous as such selection produced cumulative change. This evolution would continue until the inequality between the sexes disappeared. The 1.00 sex

ADAPTATIONS OF THE GENETIC SYSTEM

ratio would be a point of stable equilibrium at which no individual would have an advantage merely because of his sex. We would expect this 1.00 sex ratio to prevail in strictly sexual populations to the extent that natural selection, on the basis of individual reproductive success, is the determining factor.

The two sexes often have different mortality rates at various stages of development. This raises a problem as to what stage of development should be characterized by the 1.00 sex ratio. If the sexes are equal in number at conception, but the males are only half as likely as females to survive to adulthood, there would be twice as many adult females as males. This would make the average adult male twice as successful as the average adult female. Sex, however, is determined in the zygote, and a female zygote in this hypothetical population has twice as much chance of realizing her reproductive role as the male has of realizing his. At the zygote stage the prospects would be similar, because the immediate disadvantage of maleness would balance its ultimate advantage. It should therefore be the zygote stage that would have the sexes equalized by genic selection.

There are a number of complicating factors that alter the simple picture indicated above. The most important of these is the period of dependence after conception. From the standpoint of the zygote, it makes no difference at what point during development mortality occurs. The important consideration is the probability of surviving to a ripe old age. However, ontogenetic distribution of mortality does make a difference to the parent. If the difference in male and female mortality indicated above were entirely prenatal, an individual female that conceived only

ADAPTATIONS OF THE GENETIC SYSTEM

males would have an advantage. She would be burdened with fewer young to raise to the stage of independence, but because of the adult advantage of maleness, her smaller number of young would be expected to have as many descendants as a more numerous brood of mixed sexes. Hence, there would be an advantage, not in being a male, but in giving birth to males. This factor would cause the evolution of an increased number of males in the population until there was a 1.00 sex ratio at the stage at which the young become independent of the parents. I assume that for the human population this would be in late adolescence, and that we should expect male and female teenagers to be about equal in number. Males are slightly in the majority at birth, and perhaps greatly in the majority at conception, but they have higher childhood mortality rates and become a minority as adults in civilized countries. There is no reason to doubt that the same tendencies prevailed under primitive conditions and that the human population conforms closely to what is expected from Fisher's theory. Comparable data for other species are difficult to obtain, especially for those in which independence from parents occurs long before sex becomes apparent.

Despite the difficulty of obtaining precise and reliable data, the general answer should be abundantly clear. In all well-studied animals of obligate sexuality, such as man, the fruit fly, and farm animals, a sex ratio close to one is apparent at most stages of development in most populations. Close conformity with the theory is certainly the rule, and there is no convincing evidence that sex ratios ever behave as a biotic adaptation. Andersen (1961) reviewed the evidence bearing on a possible density regulation of populations by adjustments of sex ratio. Evidence was negative for all organisms of male heterogamety. In only one species group (*Drosophila pseudoobscura* and some close relatives) did population density seem to have any effect on sex ratio, and the effect here was to increase the proportion of females at high population density, the opposite of what would be expected of a mechanism for density regulation. In some other insects there were demonstrable tendencies for females to be more abundant at low densities. Some of these were parthenogenetic and will be discussed later. Others had sex ratio determined, as in the honey bee, by the option of the female parent, in a way readily interpretable as an adaptation for her reproductive success. Other examples involved greater mortality rates for females when developing under such stresses as might accompany artificially high population densities. Certainly none of these phenomena gives the appearance of an adaptive adjustment to the needs of populations. I conclude that there is no evidence from data on sex ratios to support the concept of biotic adaptation.

We can also conclude from Fisher's theory that sex ratio is not an organic adaptation either, in any population in which the sexes are numerically equal at the stage of initial independence from parents. Sex ratio provides the curious situation in which natural selection, instead of producing an adaptation, abolishes its function. In a hypothetical human population in which teenage boys are twice as numerous as their girl friends, any person with a tendency to have nothing but daughters would be in possession of an adaptation that would give him an important evolu-

ADAPTATIONS OF THE GENETIC SYSTEM

tionary advantage. As this advantage became more prevalent, as a result of natural selection, it would become less and less of an advantage. When it became so common that the adolescent sex ratio was unity, it would completely lose its adaptive significance, and would be a selectively neutral character. Sex ratio has significance as an organic adaptation only in a population that is out of equilibrium with itself, and this is probably an uncommon state. In most populations, neither an individual's sex nor the sex ratio of his offspring has any influence on his reproductive success.

Nevertheless, it does appear as if each individual were designed to produce equal numbers of male and female offspring as a long-term average. The heterogametic sex usually produces male-determining and female-determining gametes in nearly equal numbers. The opposite sex produces uncommitted gametes. For both sexes, this seems like an effective adaptation for producing a 1.00 sex ratio in offspring, and it is tempting to attribute this machinery to selection for that effect. Perhaps it can be explained merely as an incidental consequence of the evolution of the sex chromosomes themselves. The only way I can conceive of the 1.00 sex ratio as being an organic adaptation is to relate it to the primary function of sexual reproduction. If the maximization of genetic diversity among the offspring is the primary function of sexuality, as I argued earlier, the 1.00 sex ratio would contribute to this effect. If the difference between male and female is thought of as a quantitative variation —perhaps a measurable degree of maleness—having equal numbers of sons and daughters would be the condition of maximum variance in this quantity. The precise segregation of the sex chromosomes would thereby have the same significance as the precise segregation of the autosomal genes. The basic outline of this theory of population sex ratio as a function of the advantage of being of the minority sex and the advantage of increasing the number of the minority sex in one's progeny was provided by Fisher (1930). Detailed recent treatments were provided by Bodmer and Edwards (1960) and by Shaw (1958). Edwards (1960) neatly summarized the problem of whether sex ratio is the result of organic or biotic adaptation. Lewis and Crowe (1956) reviewed the problem of the natural selection of sex ratio in dioecious and in partly male-sterile plant populations.

There are some examples of environmental sex determination that can be thought to illustrate organic adaptation. It would be an advantage to an individual, before its sex is irrevocably determined, to have some indication of the sex ratio in its adult habitat, and to decide its own sex on the basis of such information. This is what happens in the echiuroid *Bonellia* (Borradaile et al., 1961). The larva is sexually neutral. If it settles in an area where females are abundant, it is likely to come in contact with one, and when this happens it attaches. Such attached individuals develop into males, parasitic on the female, and the females' eggs are fertilized by such parasitic males. If no female is contacted by the larva, as is likely to happen where females are scarce, it develops into a female. Thus each individual adjusts its sex to the opportunities presented by its demographic environment.

Another example of such adjustment to demographic environment is provided by the dioecious plant *Melandrium* (Correns, 1927). When a female is in an area of abundant males, its stigmas will receive a great abundance of pollen. It then produces a great overabundance of female seeds, by a differential rate of growth of the male-determining and female-determining pollen tubes in its styles. The environment is such as to place a premium on daughters. But if only a few pollen grains fall on the stigmas, an indication of a local scarcity of males, the proportion of male seeds increases and may approach a half of the total. In this way the plant acts to optimize the sex ratio of its progeny in relation to their probable demographic environment. It is possible that this is a common adaptation of dioecious plants (Lewis, 1942). Lewis speculated that male heterogamety, which is the common condition in dioecious plants, is adaptive in that it provides for the control of the sex ratio of offspring in the manner demonstrated for *Melandrium*.

THE FOREGOING examples are all of populations in which it takes a male and a female to make a new individual. We would expect a different outcome of selection in a species in which a female can produce viable offspring parthenogenetically. This capacity would be especially important in what have been termed fugitive species (Slobodkin, 1962), which frequently invade new habitats with an extremely small number of colonists. Sampling error would frequently make such colonists unable to reproduce sexually because the colonists might all be of the same sex. A male would have no way of escaping this difficulty, but a female, with a capacity to reproduce

ADAPTATIONS OF THE GENETIC SYSTEM

parthenogenetically, could initiate a population all by herself and achieve an enormous degree of reproductive success. This advantage was pointed out by Stalker (1956), and it would have the effect of directing the evolution of populations in which the vagile individuals would all be parthenogenetic females. The advantage of being such a female would persist after successful colonization through as many generations as parthenogenesis remained an effective method of reproduction, and selection would favor genes that increased the likelihood of becoming a female under such conditions.

This, of course, is exactly what we find. In such animals as aphids and cladocerans the dispersal stages are always females, and the proportion of females may approach unity during a long period of parthenogenetic reproduction. With the advent of conditions that favor sexual reproduction, a matter discussed earlier in this chapter, the picture changes rapidly. The females become physiologically and behaviorally readjusted for sexual reproduction, and maleness, by its scarcity and the female abundance, becomes advantageous. Males are produced in large numbers whenever changed conditions make sexual reproduction adaptive.

I would conclude that sex-determining mechanisms, like other aspects of the genetic system, are often interpretable as organic adaptations designed to optimize the sex ratio of a group of offspring and clearly show the effects of the selection of alternative alleles on the basis of the reproductive fitness of parents. Respectable evidence for biotic adaptation is nowhere to be seen. The general acceptance of this conclusion for the various phenomena discussed in

ADAPTATIONS OF THE GENETIC SYSTEM

this chapter would be a significant break with tradition, because the mechanisms of the genetic system have regularly been interpreted as biotic adaptations. Even R. A. Fisher, who generally opposed any recognition of biotic adaptation, was inclined to regard sex as the one exceptional adaptation that was (1930, Chap. 2) "evolved for the specific rather than for the individual advantage."

CHAPTER 6

Reproductive Physiology and Behavior

AN INDIVIDUAL is fit if its adaptations are such as to make it likely to contribute a more than average number of genes to future generations. Fitness may be defined as "effective design for reproductive survival." One of the more provocative alternative definitions is Medawar's (1960, p. 108):

> The genetical usage of "fitness" is an extreme attenuation of the ordinary usage: it is, in effect, a system of pricing the endowments of organisms in the currency of offspring, i.e., in terms of net reproductive performance. It is a genetic valuation of goods, not a statement about their nature or quality.

The method of measuring fitness would be to measure net reproductive performance. One might get such a measure of a past individual by counting its present descendants and comparing this count with the mean for the descendants of the individual's contemporaries. In this way we could tell how fit a past member of a population was, but, lacking clairvoyance, we could not tell how fit a present member is. I would find this something of a handicap in discussing adaptation.

A more serious objection to Medawar's definition would be that it implies that an important component of fitness is luck. Some genotypes may be more likely to survive than others, but individuals differ not only in genotype, they differ in the whole complex of moment-to-moment circumstances in which they find themselves. This relationship of fitness, chance, and survival was discussed in detail in Chapter 2. My main criticism of Medawar's statement is that it focuses attention on the rather trivial problem of the degree to which an organism actually achieves reproductive survival. The central biological problem is not survival as such, but design for survival.

Medawar's definition has great value, however, in its focusing attention on reproductive success as the only possible basis for the operation of natural selection. Simple and obvious as this point is, it has apparently been missed by some prominent and able biologists. For example, both Cole (1954) and Emerson (1960) have stated that one cannot explain the origin of mammary glands on the basis of selection for individual fitness because they contribute to the nutrition of another individual. The mammary glands are directly involved in a female mammal's attempts to increase her "currency of offspring." What could be a clearer example of a character that contributes to individual fitness?

It is possible and often convenient to recognize two categories of adaptation, those that relate to the continued existence of the individual soma, and those involved in reproduction. Basically, however, all adaptation must relate to reproduction. Somatic survival is favored by selection only when the soma is necessary to reproductive survival. Heart failure and mammary failure have exactly equivalent effects on the fitness of a female mammal (except in the one bottle-manufacturing species). Similarly all the adaptations of embryonic, larval, juvenile, and adult forms have meaning only as mechanisms that promote the survival of the genes that direct the morphogenetic sequence.

In this chapter I will confine attention to adaptations that relate directly to reproduction. Such processes are of crucial importance to the subject of this book. Somatic mechanisms, such as those relating to respiration, nutrition, etc., are usually said to be organic adaptations for individual survival, but reproductive processes are said to relate to the survival of the species. Certainly species survival is one result of reproduction. This fact, however, does not constitute evidence that species survival is a function of reproduction. If reproduction is entirely explainable on the basis of adaptation for individual genetic survival, species survival would have to be considered merely an incidental effect. The issue must be decided on an examination of the mechanisms of reproduction to determine whether they show design for group survival or for the survival of the genes of the reproducing individual.

Design for individual genetic survival is easily illustrated by the interactions between a parent and its offspring. Such interactions can be regarded as adaptations directly concerned with the reproductive survival of the parent and the somatic survival (for later reproduction) of the young, and there is no need to postulate an additional function of survival of the species. If this is the correct interpretation it should be possible to show that every adaptation is calculated to maximize the reproductive success of the individual, relative to other individuals, regardless of what effect this maximization has on the population. Every important aspect of the life cycle should make an efficient contribution to this goal. The adaptations by which individuals attempt to reproduce should never show any compromise with considerations of group welfare. There should be no instances of organized teamwork involving individuals that are not members of the same family. Any such compromises or cooperative groups would have to be interpreted as biotic adaptation.

AT THE outset it must be realized that the maximization of individual reproductive success will seldom be achieved by unbridled fecundity. Even an adult tapeworm uses some of its resources for its own vegetative needs. If it were to sacrifice all of its somatic tissues for the production of gametes it might increase today's production of offspring, but it would lose all that it would have produced tomorrow and on succeeding days. Only if it had no chance of surviving until tomorrow would it be of advantage to spend all of its resources today. The tapeworm can illustrate another principle as well as any other organism. With the same expenditure of material it could double its fecundity merely by halving the average size of the eggs (actually encased, dormant larvae) released into the intestine of its host. If such a curtailment of the material resources provided to each larva reduced the average survival by more than half, however, the increased fecundity would mean reduced reproductive success. I presume that egg size in tapeworms conforms to some optimum compromise between the advantages of high fecundity and of adequately providing for each of the young.

A very illuminating example of this principle of the optimization of the number of zygotes in relation to the resources provided for each was discussed by Lack (1954B). He showed, for several species of birds, that an above average clutch size usually resulted in a below average number of successfully fledged young. Similarly a below average clutch size resulted in a below average number of survivors, although the probability of survival of a given zygote might be higher. The population mean was that which resulted in the greatest number of successful fledglings under normal conditions. That this average fecundity was considerably below the number of zygotes that could potentially be produced by each pair is apparent from the fact that when eggs are lost they are readily replaced. Apparently fecundity was optimized according to the number of young for which the parents were able to find sufficient food. Above average clutch sizes resulted in a decrease in the nutrition of each nestling to the point at which the total survival was less than if there had been fewer nestlings to start with.

As Lack pointed out, there has been no end of uncritical thinking based on the assumption that reproduction is adjusted to the needs of populations. He referred to the frequently encountered statements to the effect that in species with high juvenile mortality rates the parents are very prolific in order to compensate for the losses. Conversely we are told, species with low mortality rates in the young stages do not need to be prolific, and have correspondingly low fecundity.

There is certainly a correlation between fecundity and mortality rate, but I would support Lack's interpretation that the cause-effect relationship is the opposite to that usually recognized. Species have high juvenile mortality as a result of their great fecundity. Conversely, low mortality rates are a result of low fecundity.

There are two factors in this dependence of mortality on fecundity. First of all, no matter what the school of thought to which one subscribes on the factors that determine population size, it is obvious that no species, no matter how well adapted and successful, can continue to increase indefinitely in a limited environment. Sooner or later, the carrying capacity of the environment will be reached or some other factor will check the increase. Once this condition is reached or even approached, a high fecundity merely increases the number of individuals that must die without issue.

The second factor is that a parent or a pair of parents has limited resources. It can provide only so much in the way of materials and efforts on behalf of a group of young. If the young are few, it can provide each with a large share of food, protection, or other advantages. If the young are many, it can provide each with only very limited resources for meeting the dangers and stresses to which it will be subjected. Such differences in initial advantages inevitably result in different mortality rates. Lack's work provides a good example of how decreased resources per zygote result in increased mortality rates, and the phenomenon is apparently of general significance. Lack found indications of its applicability in a number of other groups. In mammals, for example, he found that death rates increased as litter sizes increased, and a maximization of reproductive success with below maximum litter size was found in at least one species. He also pointed out that many animals can adjust fecundity to food supply. This ability is most marked in the social insects.

Some fish eggs, notably those of elasmobranchs, contain several grams of yolk reserves. Other fish eggs weigh less than a milligram. It is not hard to understand why the smaller eggs and the minute larvae that they produce should have greater mortality rates than the larger eggs. Parallel phenomena are found among plants. The difference in mortality rate between coconuts and beechnuts must be largely attributable to differences in initial resources.

THE LIMITATIONS of fecundity considered above all relate to situations in which a certain amount of nourishment is provided to the young by the parents, and this nourishment is divided up among the optimum number of young. Providing food, however, is not the only kind of service performed by parents. In the gallinaceous birds, for instance, food is provided by the mother in the form of yolk, but the young start feeding themselves as soon as they hatch. The clutch sizes would not be limited by selection in the manner envisioned by Lack for altricial birds. The greater fecundity of the gallinaceous birds is interpreted by Emerson (1960) to be an adaptive adjustment of reproductive rate to the increased mortality that results from living and nesting on the ground. I would interpret the increased mortality as an ecologically inevitable consequence of the increased fecundity. The large clutch sizes of gallinaceous birds can be attributed directly to the lack of any need for the parents to feed the precocial young. I would expect clutch size to be limited partly by the necessity for each egg to be fairly large if it is to produce a precocial chick,

REPRODUCTIVE PHYSIOLOGY & BEHAVIOR

partly by the fact that it would be wasteful to produce more eggs than can fit under one brooding hen, and mainly by an optimization of reproductive effort for the parent, as discussed below.

The tendency to think of reproduction as a biotic adaptation has so clouded the thinking of biologists as to cause the acceptance of a major generalization without any respectable evidence. I refer to the supposed relationship between fecundity and parental care in fishes. The literature is full of statements (e.g., Lagler, Bardach, and Miller, 1962) to the effect that since the sunfishes guard their eggs in a nest and thereby lower their embryonic mortality rate, they do not need to produce enormous numbers of offspring and have correspondingly reduced their fecundity to a few thousand eggs per pair. By contrast, it is said, those species that do not guard their eggs, such as the cod and the halibut, must make up for this deficiency by producing enormous numbers of eggs. The supposed "evidence" that has been cited always involves small fishes to exemplify those with parental care and large ones to represent those without. If one were to compare a channel catfish (large fish with parental care) with a sand shiner (small fish without) one would find "evidence" for the opposite relationship.

If suitable adjustments are made for differences in the size of the fishes and in the size of the eggs, one finds no evidence whatever that parental care and fecundity are in any way dependent (Williams, 1959). Fecundity in fishes is apparently limited by food supply, the duration of the favorable spawning season, and, ultimately, by the amount of space available for the storage of eggs in the body of a female.

The only way in which natural selection might reduce fecundity in relation to parental care is through an advantage in decreasing the expenditure of materials for the production of gametes so that the parent would be better able to care for the eggs and young immediately afterwards. In most groups, however, it is only the female that expends a major part of her resources in the production of gametes. The material contribution of the male is slight, and it is the male that assumes complete responsibility for the eggs after they are fertilized. A reduction in fecundity in such species would not help with the care of the young. Only in those few groups, such as the cichlids, in which the female shares in the care of the young, and sometimes bears most of the responsibility, would I expect to find a reduction in fecundity.

The problem of the optimum number of young is related to that of the optimum mass of young, which will be discussed below. If we assume a certain optimum mass, however, there would be the problem of how this mass should be divided up so as to maximize the number of surviving offspring. If 200 grams of material is the optimum expenditure for a 1,000-gram female fish, would her reproductive interests be best served by laying 200,000 one-milligram eggs, or 2,000 hundred-milligram eggs, or by the birth of 20 ten-gram young? I assume that the production of a 200-gram mass of eggs would require about the same material expenditure regardless of the size of the individual eggs, but that giving birth to 200 grams of advanced young would imply a much greater sacrifice. Such young would necessarily metabolize for a long time in the mother and have to be provided with food and oxygen greatly in excess of that present at birth.

These two problems, that of the optimum amount of effort, and that of its most effective distribution, can be considered separately. I will treat the distribution problem first, and consider whether a fish will succeed better by laying 200,000 one-milligram eggs or a smaller number of larger eggs. This problem has been explored by Hubbs (1958) and by Svärdson (1949). A major factor is undoubtedly the conditions that normally prevail with respect to potential food supply for the young, and the sizes of eggs that would be most liable to predation or other loss. If, at the time of spawning, food for the young will continue to be in low supply for some time, I would expect the eggs to be large and yolky, and therefore slow-developing and few. The same adaptation might follow from the presence of predators that are very effective against, perhaps, one-milligram eggs, but which are rather ineffective against eggs of several times that size. On the other hand, if there were normally an abundance of food suitable for a fish hatching from a one-milligram egg, and no special predator pressures against such eggs and the ensuing minute larvae, I would expect the species to evolve and maintain an egg of that small size. For such a species the optimum fecundity would be at a high level. Gotto (1962) reasoned that the problem of number vs. size in parasitic copepods may be determined by the ecological problems faced by the young, such as the probability of finding a host. He also pointed out a number of promising lines of investigation on the natural selection of egg number and size.

REPRODUCTIVE PHYSIOLOGY & BEHAVIOR

The maximization of reproductive success requires that the eggs be of such a size and laid in such circumstances that the effective growth of this investment is as rapid as possible. The effective growth is the total for the component individuals minus the decrement from mortality. The number of individuals in the group of offspring steadily decreases, and unless those that survive grow fast enough to bring about an increase in the aggregate mass, the parental investment will be poorly represented in the offspring generation. This must always be true in any population that is not changing rapidly in size.

This relation between growth, mortality, and reproductive success can be illustrated by a hypothetical fish that weighs 100 grams at one year of age, at which time it spawns 10 grams of eggs and then dies. For a pair of parents to have average reproductive success, this 10 grams of eggs must grow in one year to two individuals representing a mass of 200 grams. It does not matter whether these are two out of an original three, or two out of a thousand, nor does it matter whether mortality or growth is mainly early in the year or late in the year. Only the end result counts.

For reasons explained in Chapter 3, growth will be most hurried during periods of heavy mortality, and in most species the early stages will suffer the heaviest mortality. Perhaps this reverse relation between mortality and growth rate would be such that the aggregate growth of the group would be more or less linear through the year. This cannot be true, however, for the egg stage. Eggs must suffer a steady mortality, but cannot grow by assimilating food from their environments. First they must perform the

REPRODUCTIVE PHYSIOLOGY & BEHAVIOR

herculean morphogenetic task of providing themselves with effective digestive systems and the complex sensory and motor patterns necessary for feeding. A spawning will always diminish in mass before it can begin to grow. The situation is diagramed in Figure 4, which traces the one-year history of

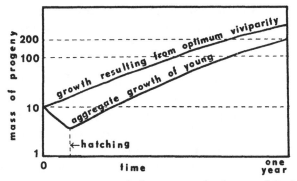

FIGURE 4. Optimization of the stage at birth in a fish that acquires internal fertilization.

changes in mass of a hypothetical 10-gram spawning of a thousand eggs. It is apparent from the diagram that the mother would be more successful if she spawned, not 10 grams of eggs, but 10 grams of young at the stage at which they start an aggregate increase in mass. Her offspring would then, *ceteris paribus*, follow a course parallel to the actual one but higher at every stage (upper line in the diagram). This advantage, however, cannot be realized unless the preadaptation of internal fertilization is present. Without internal fertilization the fish must shed her young at the gamete stage in order to achieve their fertilization. The resulting zygote must necessarily be extremely vulnerable to predators and incapable of assimilating food for some time. Her investment in posterity must depreciate before it can grow.

If internal fertilization has been evolved, there will immediately be selection for the retention of the young until the moment at which they would, on release, gain more by their aggregate growth than they would lose by death. This explains the frequency in fishes and most other groups of animals, of an association between internal fertilization and viviparity. Selection would favor the retention of zygotes until they reach a stage in development determined by the demographic factors of growth and death rates of the young. This change could be called the optimum shift towards viviparity. It would be influenced, of course, by the greater material sacrifices required of the mother for producing advanced juveniles, compared with what would be needed to produce the same mass of eggs.

There are two outstanding exceptions to the association of viviparity with internal fertilization, the birds and the insects. I find the absence of viviparity among the birds most mysterious. The enormous egg mortality rates in most species (Lack 1954A) suggests that a bird could benefit greatly from viviparity, even if fecundity were greatly reduced so that the weight of the foetuses would not seriously interfere with flight. In view of the great variation in breeding habits and population structures I would certainly imagine that viviparity would be favored in at least some species. Perhaps the birds are lacking some important preadaptation, which early mammals possessed, for overcoming the immunological obstacles to viviparity.

No such excuse can be offered for the insects. It

is quite apparent that insects can become viviparous, because some of them have done so, and they represent taxonomically diverse groups (Hagan, 1951). About half of the major orders include some viviparous groups, although the relative number of viviparous genera and species is small. Perhaps the scarcity of viviparity among the insects is due to the reduction of fecundity that must accompany the carrying of young to an advanced stage in their development. This is certainly true of the social insects. If a queen bee had to house and nourish every one of her offspring within her own body until they reached a larval or even an advanced embryonic stage, she could not be nearly so fecund as she is.

The primary advantage of viviparity is that it greatly reduces the losses that normally afflict the earliest stages of development. The insects, especially the social insects, characteristically lay their eggs in protected niches, in which mortality risks may often lie below what the adults themselves experience. This factor may remove the selective advantage of viviparity. If this is the true explanation, it should be possible to show that those insects that have become viviparous are those that have life histories that reduce the availability or effectiveness of protected niches for the egg stage.

THE DISCUSSION to this point has assumed that there is some optimum expenditure of materials on the part of the mother. The problem considered was how the optimum use of the material could be attained. Nothing has been said of what determines how much material should be expended. This is a part of the larger problem of what determines how much effort an organism should expend, and to how much danger it should expose itself, in its efforts to reproduce. It should be obvious that parental sacrifices are sometimes enormous and sometimes very slight. In dealing with this problem I will take it as a basic axiom that selection will adjust the amount of immediate reproductive effort in such a way that the cost in physiological stress and personal hazard will be justified by the probability of success. In most species an individual's soma is its prime resource in the struggle for genetic survival. It is an investment that must not be vainly jeopardized. A well-adapted individual is one that will engage in reproductive activities only when the chance of success is at some peak value and the probable costs and hazards at some low point.

The possibility of an objective assessment of physiological cost is indicated by the work of Barnes (1962) and of Crisp and Patel (1961). They were able to measure the effect of spawning on the growth of barnacles and showed that spawning was performed at the expense of an appreciable amount of growth. Such curtailed growth, in turn, would mean lowered fecundity in the next breeding season. I know of no work on the effect of reproduction on mortality rates. Such an effect might be sought in field studies of nesting birds. Those few species in which individuals more or less regularly breed in alternate years would be especially favorable material.

The relative value of the "somatic investment," compared with that of the immediate reproductive opportunity, will be determined by the demographic environment. The relationship can be made clear by a hypothetical example. Suppose a species breeds once a year, in the spring. This seasonal limitation would result from the probability that reproductive success would be at a peak value relative to cost during the spring. Suppose further that the breeding habits of the species are very simple; all it has to do is produce gametes and release them at the proper moment. Suppose also that each individual grows in size and therefore in potential fecundity all its life. Now imagine two young individuals approaching the stage at which the decision will be made as to whether to breed or not to breed in the coming spring. The two are identical in every way (size, age, health, nutrition, genotype) except that one has a gene that says, "Yes—start making gametes for next spring," and the other has an alternative allele that says, "No—wait another year." Which individual (and gene) is the fittest? The answer all depends on the demographic environment. The survival value of immediate reproduction would be increased by a high annual mortality rate (so high that the coming spring might well be the last chance) and a low yearly increase in fecundity. The survival value of the delay would be greater if there is a low annual mortality rate and high annual fecundity increase, which would greatly reduce the value of the immediate reproductive opportunity in comparison with prospects for the more distant future. If such things as annual mortality rates are variable for the population, the long-term mean would be the decisive factor. Where any important ecological factor is variable, the fittest gene would be one that would say, in response to the question on whether or not to undertake reproduction, "Yes, if . . . ," with an appropriate *if* clause. The favorable selection of such "yes-if" genes would make reproduction, or its intensity, a function of such time-dependent variables as the health and nutrition of the individual, the weather, and other factors that would influence either the probability of successful reproduction or the extent of the somatic sacrifice that would be necessary.

If the demographic factors of age-dependent fecundity and of adult mortality rates are the determinants of reproductive effort, as envisioned above, certain generalizations are expected to follow. Those species characterized by a long adult life during which they grow continuously in fecundity should have a low level of reproductive effort in any given breeding season. On the other hand, those species with determinate growth and a heavy mortality from one breeding season to the next should expend great efforts and often take great risks in their efforts to reproduce. The extreme condition would be found in species that can breed only once. In them we would expect the most extreme examples of the risk of life and of physiological expenditures in the reproductive process. There are a number of relevant examples among the fishes. The Pacific salmons spawn only once, and among them we find the expected emphasis on reproductive functions to the detriment of the parental somata. Some of them undertake the longest migrations known in the salmonidae or any other group of anadromous fishes. In the preparation for spawning the digestive system atrophies so as to make continued existence impossible, but this atrophy supplies materials and space for gametes, and unburdens the fish of extra weight, unnecessary to the single reproductive effort, in its upstream journey. The mouth of the male undergoes changes which aid the fish in sexual combat, but make it unfit for the efficient ingestion of food. The belligerence of the males in competing for territories and females is unmatched in any other fish of comparable size. These and other extremes of reproductive effort are clearly adaptive in a fish for which the one opportunity will be its last. The origin of semelparous life cycles is one aspect of the evolution of aging (Williams, 1957:408).

There are other such life cycles among fishes. The lampreys are one example, and they parallel the Pacific salmons in their reproductive behavior and physiology. In tropical America and perhaps elsewhere there are annual fishes, species that grow to maturity and reproduce in temporary ponds that disappear in the dry season. Before the pond dries up, however, they spawn drought-resistant eggs that lie dormant until the rains again fill the ponds. These then hatch and grow to maturity before the next dry season. Myers (1952) has provided a general account of these fishes. The whole group is characterized by a high degree of sex dimorphism and very bright colors in the males. The males are also extremely pugnacious in their relations with competing males. In both pugnacity and coloration they are apparently more extreme than their perennial relatives, and thereby confirm the expectations expressed here. The measurement of these factors is, of course, subject to the uncertainties of subjective judgment and of the unfortunately meagre amount of information that we have on these fishes.

Organisms that breed only once constitute, as indicated above, one extreme of a spectrum. At the other extreme are species that have very low mortality rates as adults, often live for many breeding seasons, and increase in fecundity with each passing year. In between there is a broad range of life-cycle types that would provide a broad range in the values of immediate reproductive opportunities, compared to the total opportunity inherent in the future prospects of the individual. In each species the level of reproductive effort should always reflect the ratio of these two quantities. Formally the expectation is

$$E = \frac{F_0}{\sum_{i=1}^{\infty} (F_i S_i)}$$

where E is the measure of optimum reproductive effort, in terms of physiological stress and risk of life, F_i is the most probable fecundity, or other measure of reproductive effectiveness, in the ith breeding season (the current or impending breeding season would be number zero) and S_i is the probability of surviving until the ith breeding season.

This ratio is undoubtedly subject to extreme variation. There are fishes that show a steady increase in fecundity for many years. In many cases a simple linear increase would be a fair approximation for a number of years. Such a fish might lay about 1,000 eggs in its first spawning season, 2,000 in its second, and so on. A more or less steady year-to-year adult survival rate of 0.8 would probably be fairly accurate in some species, and greater rates of survival probably occur. The value of E for such a species would be only about 4 per cent of that for a species with constant adult fecundity and a survival probability of only 0.5 per year. More extreme differences no doubt occur. Note that the value of the denominator would be zero in a species that dies after breeding once. In such species selection would, as indicated above, favor the expenditure of all available resources in the one effort.

The fishes would be an excellent group in which to test a theory of this sort because of the extreme range of breeding behavior and population structure, but unfortunately we have few reliable indications of E, F, and S for natural populations. Much of the information that we do have relates to commercially exploited species, which now may have population structures very different from those that have normally prevailed. Nevertheless, I think that there are some indications that the fishes do conform to expectations. These indications derive from the way in which E, F, and S relate to body size.

With respect to the factor F, it would seem that determinate growth, which implies nearly constant fecundity once maturity is reached, is known in only a few fishes, all quite small (Wellensiek, 1953). Many other small fishes, such as those commonly kept in aquaria, grow very little after reaching maturity. Some statistics that I have gathered indicate that fishes that never reach more than 50 millimeters will normally reach maturity at about one-half to three-fourths maximum size. Fishes that commonly exceed 500 millimeters, however, usually attain maturity at between one-fifth and one-half maximum size. It seems to be a general rule that large fishes have a more indeterminate growth pattern than small ones. This should mean a greater reproductive effort among the smaller species.

It is a common expectation that large organisms have lower mortality rates than small ones. This expectation is confirmed in fishery statistics, which seldom list age groups of beyond three or four years for fishes of a few centimeters maximum length, but often provide data on fifteen- or twenty-year life spans in nature for such large species as sturgeons, gar-pikes and halibut. Here again is a factor that should raise the reproductive efforts of small species relative to the large ones.

The next problem is to decide whether small species do, as a general rule, put forth a greater effort in reproduction than large ones. There is no way of obtaining a numerical measure of reproductive effort, but it is possible to rank different kinds of breeding habits according to which involves the greater and which the lesser effort and sacrifice. We can assume that the effort or hazard is (1) greater when a species spawns a large mass of eggs, relative to the fish that produces them, than when only a small mass is produced, (2) greater if a species spawns several times in a breeding season than if it spawns but once a year, (3) greater if there is a conspicuous breeding coloration than if the fish maintains normal coloration through the breeding season—and we would assume that the more conspicuous the color, the greater is the somatic hazard, (4) greater in species that engage in conspicuous courtship displays than in those with very simple pre-spawning behavior, (5) greater in species with highly competitive relations among members of the same sex, especially when the competition includes actual fighting with effective weapons, (6) greater in species that maintain territories than in nonterritorial species, (7) greater in species that provide some special protection for their eggs, such as laying them in protected niches or guarding them in nests, than in species that merely scatter their eggs about, (8) greater in viviparous species than in oviparous species.

I am confident that if a poll were taken among ichthyologists, nearly all would agree that with respect to most of the criteria above, small fishes usually show greater reproductive effort than large ones, although all would be able to think of some outstanding exceptions. The first point listed requires some comment. In a preliminary survey of the literature on relative ovary weights (Williams, 1959), there were some indications of relatively larger egg masses in smaller species. Extensive data for a tuna showed that none ever produced egg masses in excess of 5 per cent of the parental body mass, whereas relative ovary weights of more than 20 per cent were common in very small species. I believe it is safe to say that one never sees such extreme body distortion by egg masses in large species as are found among such small fishes as darters and sticklebacks.

I have gathered evidence on each of the other points, but proper presentation and discussion would be too lengthy for treatment here. For present purposes I will rely on the assumption that ichthyologists would find it in accordance with their experience to recognize a size bias in each of the points enumerated, with the greater effort being characteristic of smaller species in each case. I have found this to be true of the ichthyologists I have consulted, even for the last two points, on which I believe the evidence to be quite equivocal, or even contrary to the expectation of greater effort among smaller species. There are many quite small fishes, such as the sticklebacks and the gouramis, which do place their eggs in elaborate nests and guard them from possible predation, and these are the species that an ichthyologist first thinks of when egg-guarding behavior is mentioned. Nevertheless, there is a sizable proportion of large species with comparable habits. The black basses, bowfin, and larger catfishes comprise a fair proportion of the larger fishes of the United States, and all guard their eggs in a nest. Many other large freshwater fishes do the same, and the habit seems especially prevalent among the more archaic types, such as *Protopterus* and *Arapaima*. In at least one family of nest-builders, the Centrarchidae, the larger species show better-developed parental care than the smaller (Breder, 1936).

Oral incubation is another kind of special egg protection that shows no clear dependence on size. The smallest mouth-breeder is apparently *Betta brederi*, which reaches a length of about 70 millimeters. The mouth-breeding cichlids and apogonids can be thought of as average-size fishes; the mouth-breeding catfishes, rather large ones. Much the same could be said for viviparity. Advanced, placental viviparity is known in several families of quite small freshwater fishes, most of them closely related, in the intermediate-size marine embioticids, and in the very large sharks.

As a summary of the evidence I would say that expectations are, in general, confirmed, and that small species usually show greater reproductive effort than large ones, but that exceptions may have to be recognized for those aspects of augmented reproductive effort that involve giving special protection to the young. A tentative explanation for the absence of a clear size dependence in the incidence of parental defense of eggs in nests is that, while a small fish may have a greater need for such protection for its young, its small size may make it less effective in providing such protection. Oral incubation and viviparity would normally require some reduction in fecundity. This circumstance may tend to reduce their value for small fishes.

The birds are another group of animals in which the expected relation between size, or other indications of mortality rate, and reproductive effort might be tested. The desired precision of demographic information is, of course, lacking, but a confirmation of expectations is the general rule. The lowest effort seems to be expended by large marine or predatory birds, which have very low adult mortality rates. They frequently skip nesting seasons, typically have very few young, and may show little tendency to replace lost eggs. The eggs are not only few, they are small in proportion to adult size. The mature birds seldom show striking sex dichromatism or elaborate courtship behavior. Small birds and ground-living birds, however, frequently show elaborate courtship, lay relatively many and large eggs, usually replace them when they are lost, and, according to Mottram (1915), are more likely to be sexually dichromatic. The adaptive adjustment of the reproductive behavior of each individual to the probability distribution of its mortality is the obvious and parsimonious explanation for these observations. These various aspects of phylogenetic variation in the reproductive physiology and behavior of birds are reviewed by Wynne-Edwards (1962). His interpretation of them is quite different from mine.

THE OPTIMUM reproductive effort should not only vary among species, it should also change with time in any given species, at least in those species that increase appreciably in fecundity or other measure of reproductive performance for several seasons after reaching sexual maturity. A halibut in its tenth spawning season can expect some future increase in fecundity, but not nearly so much as a halibut in its first spawning season. This would make E in the effort equation an age-dependent variable. The intensity of reproductive effort should increase with age, and the greatest increase should be between the first and second spawning. It is a common observation that the first attempt at reproduction in any organism is likely to be of low intensity, and this observation accords with expectations expressed here. Whether reproductive effort continues to show slight increases with advancing age, in those species in which such increase would be expected, is a question for which the answer is less immediately clear, although there is evidence that the expectation is realized in some large marine fishes. Hodder (1963) showed that female fecundity increases at a steeper rate than individual mass for many years in a haddock population, and reviewed indications that this is also true in other species. For most female fishes, fecundity in relation to size should constitute an accurate measure of reproductive effort.

The optimization of effort, combined with the different primary roles of the sexes, may explain the different secondary roles of the sexes. It is commonly observed that males show a greater readiness for reproduction than females. This is understandable as a consequence of the greater physiological sacrifice made by females for the production of each surviving offspring. A male mammal's essential role may end with copulation, which involves a negligible expenditure of energy and materials on his part, and only a momentary lapse of attention from matters of direct concern to his safety and well-being. The situation is markedly different for the female, for which copulation may mean a commitment to a prolonged burden, in both the mechanical and physiological sense, and its many attendant stresses and dangers. Consequently the male, having little to lose in his primary reproductive role, shows an aggressive and immediate willingness to mate with as many females as may be available. If he undertakes his reproductive role and fails, he has lost very little. If he succeeds, he can be just as successful for a very minor effort as a female could be only after a major somatic sacrifice. Failure for a female mammal may mean weeks or months of wasted time. The mechanical and nutritional burden of pregnancy may mean increased vulnerability to predators, decreased disease resistance, and other dangers for a long time. Even if she successfully endures these stresses and hazards she can still fail completely if her litter is lost before weaning. Once she starts on her reproductive role she commits herself to a certain high minimum of reproductive effort. Natural selection should regulate her reproductive behavior in such a way that she will assume the burdens of reproduction only when the probability of success is at some peak value that is not likely to be exceeded.

The traditional coyness of the female is thus easily attributed to adaptive mechanisms by which she can discriminate the ideal moment and circumstances for assuming the burdens of motherhood. One of the most important circumstances is the inseminating male. It is to the female's advantage to be able to pick the most fit male available for fathering her brood. Unusually fit fathers tend to have unusually fit offspring. One of the functions of courtship would be the advertisement, by a male, of how fit he is. A male whose general health and nutrition enables him to indulge in full development of secondary sexual characters, especially courtship behavior, is likely to be reasonably fit genetically. Other important signs of fitness would be the ability to occupy a choice nesting site and a large territory, and the power to defeat or intimidate other males. In submitting only to a male with such signs of fitness a female would probably be aiding the survival of her own genes. Inevitably there is a kind of evolutionary battle of the sexes. If a male attempts to reproduce at all in a certain breeding season, it is to his advantage to pretend to be highly fit whether he is or not. If a weak and unresourceful male successfully coaxes a female to mate with him he has lost nothing, and may have successfully reproduced. It will be to the female's advantage, however, to be able to tell the males that are really fit from those that merely pretend to be. In such a population genic selection will foster a skilled salesmanship among the males and an equally well-developed sales resistance and discrimination among the females. This evolutionary effect of female choice has been formally analyzed by O'Donald (1962). The original idea, like so many other important insights, was contributed by R. A. Fisher.

The greater promiscuity of the male and greater caution and discrimination of the female is found in animals generally (Bateman, 1949). Even without such special female functions as pregnancy and lactation, it is almost always true that females contribute the greater amount of material and food energy to the next generation. A female can readily increase her reproductive effort merely by increasing the mass of gametes to the point at which further gain would not be worth the sacrifice. For males, especially in species with internal fertilization with its great economy of sperm, the problem is not so simple. A male can easily produce sperm in excess of what it would take to fertilize all the females that could conceivably be available. The reproductive effort involved in male gametogenesis would ordinarily be slight. Most of his reproductive effort can be devoted to the problem of increasing the number of females available for insemination. Hence the development of the masculine emphasis on courtship and territoriality or other forms of conflict with competing males.

An important test of this explanation is whether the expected exceptions to this difference in male and female approaches to reproduction can be demonstrated. In some species the males contribute more materials in providing for the next generation or undergo greater risks in their essential roles. The best example I know is in the pipefish-seahorse family, the Syngnathidae. In this group the females, in copulation, are not inseminated by the males. Instead, they transfer their eggs to a brood pouch in the male. There the young develop to an advanced stage with the help of a placental connection with the male blood stream. Under the circumstances, we might expect that it would be the female that would show the traditional masculine aggressiveness in courtship and general promiscuity, and the male that would show caution and discrimination. This is known to be true in some species, and it is not known to be untrue in any (Fiedler, 1954).

Parallel but less extreme modifications are seen in certain amphibians and insects. The females of the Belostomid water bugs attach their fertilized eggs to the backs of the males, who carry them until hatching (Essig, 1942). A number of tropical toads show analogous developments. In none of these groups, however, is the male's contribution of food energy or other personal sacrifice likely to be great enough to cause as marked a reversal of sexual behavior as that found in the syngnathids. The toads, however, offer a special advantage in the study of such effects. In some genera the males carry the fertilized eggs, and in others the females (Noble, 1931). A comparison of the sexual behavior of these genera would be of great interest.

There are some birds in which the male customarily assumes all or most of the burden of incubating the eggs and feeding the young. The most extreme examples are the Tinamous and Phalaropes. As expected, the female is the aggressor in courtship, more brightly colored than the male, and inclined to polyandry in both groups (Kendeigh, 1952).

The evidence strongly supports the conclusion that promiscuity, active courtship, and belligerence toward rivals are not inherent aspects of maleness. They will be developed in whichever sex cannot effectively increase its production of offspring merely by increasing its material contribution, either by an augmented production of gametes or supply of food energy to the support of young. In any species in which the two sexes show roughly similar adult mortality rates and age distributions of fecundity, there will be roughly similar amounts of effort expended by each sex in its reproductive role. If the optimum amount of effort in one sex is largely achieved in the supply of materials devoted to the next generation, that sex will do little more. If the material contributions of the other sex are minor, it will raise its efforts to the optimum level in some other way. It may spend large amounts of time in seldom successful courtship. It may spend much time and effort and expose itself to considerable danger in conflicts with competitors, and it may grow special weapons or ornamentation for these uses. These developments can be readily understood as the optimization of reproductive effort by the individuals that show them. They may contribute to group welfare, as was argued by Wynne-Edwards (1962), or may, as a number of workers have pointed out, notably Haldane (1932), seriously reduce the fitness of a species. Sexual conflict and territoriality are discussed further on pp. 238-242.

PERHAPS the best place to look for reproductive activities that show an organization for group survival would be in those species that breed in large conspecific assemblages. These reproductive groupings can be examined to see if they are adaptively organized for a collective reproductive effort or are merely statistical sums of the component individuals and their adaptations.

In one sense, all efforts in sexual reproduction, in that they are directed toward the survival of the offspring of two different individuals, are concerned with collective reproduction. A cow, in providing milk for her calf, is not only promoting the survival of her own genes, but those of some bull as well. Only half of her genes are represented in the calf. Nevertheless, sexual reproduction is the only possible kind of reproduction for a cow. To provide for the survival of her own genes, she must also provide for the survival of some competing genes. The problem of the origin of sex and the evolution of such a basically imperfect type of reproduction was treated in Chapter 5. At any rate, lactation is easily interpreted as an element in a female mammal's efforts to reproduce herself.

There are many aquatic organisms, both fishes and invertebrates, that come together in large numbers to spawn. However, I am not aware that any adaptive organization has been described in such groups, aside from the sum of the basically identical contributions of individuals. I presume that each individual joins the aggregation because there its gametes will have a better chance of fertilizing or of being fertilized than if they were shed elsewhere.

Among the gregarious mammals it is a general rule that the herds break up into pairs in the foaling season. The contribution made by males in the rearing of offspring varies from species to species. As far as I am aware, females of all species suckle their own young only. In gregarious species, in which the young of different mothers may mingle to a certain extent, there are always individual recognition factors that enable each mother to confine her solicitude to her own offspring. The truth of this point was dramatically demonstrated to me some years ago when I saw a motion-picture account of the family life of elephant seals on one of their insular rookeries. Amid the crowded but thriving family groups there was an occasional isolated pup, whose mother had deserted or been killed. These motherless young were manifestly starving and in acute distress. The human audience reacted with horror to the way these unfortunates were rejected by the hundreds of possible foster mothers all around them. It should have been abundantly clear to everyone present that the seals were designed to reproduce themselves, not their species. I know of no evidence that the adaptive organization of subhuman mammalian reproduction ever normally transcends the interactions between a pair of parents and their offspring.

Such organization has been postulated for bird colonies. Anyone who has had the unpleasant experience of being mobbed by terns can attest to the effectiveness of their joint efforts. He would acknowledge, however, that *mob*, rather than *team* or *task force*, is the appropriate term. There is no organized strategy of attack and no division of labor. Each individual is undoubtedly bolder in the attack when it is accompanied by other terns, but this is not a factor for which biotic adaptation need be postulated.

Marine bird colonies of large or moderate size may be more productive of young than small colonies. This has been interpreted (Darling, 1938; Allee *et al.*, 1949) as evidence of at least some generalized benefits deriving from sociality as such, and perhaps of a supra-organismic organization for group survival. The origin of social nesting can be explained by assuming that the vicinity of conspecific individuals is, for one reason or another, a favorable place to build a nest. In any species of which this is true, selection will produce gregarious nesting. This should be an adequate explanation for the origin and evolution of colonial nesting groups. The low success of small groups has recently been interpreted to mean, not that large numbers cause an increased success, but that large numbers indicate the most favorable nesting sites (J. Fisher, 1954; Lack, 1954A). The birds in small colonies are not any less social, they are there because they are unable to compete for positions in the more favorable nesting sites. Small colony size and low productivity would be independent results of inferior nesting sites and inferior birds.

In social birds, as in social mammals, each pair normally provides only for its own young, and has special recognition mechanisms that make this possible in the crowded rookeries. When such birds evolve new breeding habits, such as nesting in relative isolation on cliff ledges, these parent-young recognition mechanisms, being less important in bringing parents and young together, become less effective and more easily circumvented by experimental substitutions (Cullen, 1957). In colonial birds as in the elephant seal, each pair is effectively designed to reproduce itself, not its species.

An exception to this rule of parents caring for their own young has been claimed for penguins (Allee *et al.*, 1949; Kendeigh, 1952; Murphy, 1936), and the California woodpecker (Ritter, 1938). It can sometimes be observed that there is an excess of broody adults at a penguin rookery, and that several may fight for possession of a single egg. The one that wins may conceivably not be its parent. Later, when the chicks are feeding, the number of adults on the rookery may be low because some of them are out fishing. The ones that stay behind are presumed to be babysitting for those that are gathering food, the so-called *crèche* system of communal responsibility. When the missing adults come back laden with food, they appear to give it to whichever chicks indicate the most urgent need, and seem to show no discrimination in favor of their own young. It has been assumed that penguin chicks are impossible to tell apart, even by their parents. Recent work, however, has indicated that appearances are deceiving (Budd, 1962; Penny, 1962; Richdale, 1951, 1957). Competition for eggs by broody adults results from egg loss. Parents normally incubate their own eggs and accept others only when deprived of their own. Marking experiments have shown that when adults return from the sea they recognize their young by vocal signals, and avoid feeding any but their own offspring.

There are some undoubted examples of birds' helping to rear offspring that are not their own. This phenomenon may be especially marked in the California woodpecker, a bird which, for a number of reasons, is of particular interest in a search for biotic adaptation. Likewise the reproduction of the social insects is of special concern and could appropriately be discussed here. It is more convenient, however, to discuss these reproductive phenomena along with other problems of the social organization of these animals, and their treatment is deferred to the next chapter.

Early in this chapter I indicated a belief that the various aspects of the reproductive behavior and physiology of a species, its intensity, timing, ontogeny, and every important feature of its physiological and behavioral mechanisms would be precisely designed to maximize individual reproductive performance. To support this belief for even one species would be a large order. I have been able to offer only a few examples of how this approach can be used to explain and sometimes predict the phenomena of phylogenetic variation in reproductive physiology and behavior. Despite the limitations, I hope that I have been able to show that the method can shed light on some very basic phenomena, such as the male aggressiveness of most dioecious animals, and the universality of parent-offspring specificity in the relationship between the generations of gregarious species. In particular I believe that relating phylogenetic variation in reproductive functions with phylogenetic variation in the demographic factors of age-specific fecundity and death rates will reveal much of explanatory value.

CHAPTER 7

Social Adaptations

BEHAVIORAL or physiological mechanisms that operate between an individual and its own offspring are normally benign and cooperative, but interactions between unrelated individuals normally take the form of open antagonism, or, at best, a tolerant neutrality. This is usually an accurate description of relationships within a species, although there are many apparent exceptions. The prevalence of solicitude for offspring and hostility to all others is clearly shown by what are considered the higher animals. The house cat population in any neighborhood would be a good illustration.

Except for the phenomenon of general gregariousness, which will be considered later in this chapter, most of the conspicuous animal-behavior patterns between unrelated individuals are manifestly competitive. A dog would rather fight than share a bone, and if it is not hungry now it will bury the bone for its own future needs. A male robin intimidates intruders in its territory and is openly opposed to a sharing of resources with any conspecific bird except mate and offspring. Its courtship is clearly designed to appropriate an optimum share of a limited resource, potential mates. These self-seeking behavior patterns, which are widely prevalent in the animal kingdom, are easily attributed to selection for competitive efficiency in genetic survival.

SOCIAL ADAPTATIONS

Although attempts have been made (e.g., by Allee, in all citations; Wynne-Edwards, 1962) to interpret territoriality and intimidation as ultimately benign and biotically adaptive, I will assume for the moment that the basically selfish nature of such behavior is accepted by most biologists. Here I will concentrate on the less common interactions that do seem to be cooperative and benign. Although they are certainly less common, they are nevertheless quite diverse and common enough to constitute *prima facie* evidence of biotic adaptation. How could natural selection, based on the relative rates of reproduction of different individuals, favor genes that cause their bearers to expend resources to benefit their genetic competitors? This chapter attempts to answer this question, and it is my contention that one of two answers will always suffice, the choice depending on the particulars: (1) Natural selection of alternative alleles, based ultimately on the mean reproductive success of their carriers, can indeed favor cooperative interactions when they involve closely related individuals, not necessarily parent and offspring. (2) Whenever there are behavioral mechanisms by which parents aid their offspring, there will inevitably be times when such aid is provided "by mistake" to unrelated individuals.

THE MOST extreme examples of cooperative interactions are apparently of the first type. Obviously there are many examples of clonally produced "individuals" that exhibit precisely coordinated cooperative interactions: the members of a siphonophore colony, the somites of an earthworm, the cells of a metazoan.

SOCIAL ADAPTATIONS

From the standpoint of evolutionary theory, however, the concept of an "individual" implies genetic uniqueness, and this attribute is not shown by these examples. To provide benefits to a genetically identical individual is to benefit oneself. There is no paradox in one cell's committing itself to a sterile somatic role in development if this commitment benefits a genetically identical cell in the germ line. The only condition that must be met is that the benefit to the germ cell, in "the currency of offspring," must be greater than the loss of fertility by the somatic cell and the cost of its maintenance.

A problem exists, however, when there is an example of benevolent self-sacrifice among genetically different individuals. Organisms showing such behavior were termed *altruistic* by Haldane (1932). I prefer the less emotive term *social donors* (Williams and Williams, 1957). The problem is easily appreciated in considering the selection of a recessive gene d which causes its bearer to provide, at cost c, an advantage a to other members of its Mendelian population. The dominant allele D causes its bearer to refrain from such behavior. Under random mating we would have the Hardy-Weinberg zygote frequencies of

p^2 of DD
$2pq$ of Dd
q^2 of dd

After selection, these frequencies would be modified to

$p^2 (1 + q^2 a)$ of DD
$2pq (1 + q^2 a)$ of Dd
$q^2 (1 + q^2 a)(1 - c)$ of dd

SOCIAL ADAPTATIONS

The increase in each frequency by the factor $(1 + q^2a)$ is based on the assumption that the amount of benefit is directly proportional to the concentration of social donors. This concentration must inevitably decrease, however, because of the factor $(1 - c)$ by which the homozygous recessives are reduced each generation. Wright's attempt to evade this conclusion is discussed on pp. 111-112.

The outcome may be different if we assume that the social donors provide their benefits when they are young and in a group composed entirely of their brothers and sisters. The relative survival of the different genotypes in the sibling group is still the same; the donors are adversely selected. Nevertheless it is possible for the donor gene to increase its frequency in the population. Assume that the donor gene is so rare that it never becomes homozygous except in the progeny of the occasional cross $Dd \times Dd$. In this case the benefits from the presence of the donors (dd) would be enjoyed only in those rare families in which three-fourths of the individuals have the donor gene in at least heterozygous state. The benefits would be realized in groups that are not random samples of the population, but samples with atypically high frequencies of the donor gene. This association between the benefits and the genetic basis of the donor phenotype can, theoretically, counteract the disadvantage of the donors in competition with their non-donor sibs and result in an increase in the frequency of the donor gene. It appears that its continued increase is ensured by having the benefit per donor equal to several times the cost to each donor (Williams and Williams, 1957). This condition would be easily met when older and more resourceful individuals can aid their younger and smaller sibs.

This is the explanation that seems to hold the most promise for the problems posed by the social insects. From the standpoint of the main issues treated in this book there is no more important phenomenon than the organization of insect colonies, and no more important question than whether the explanation outlined above can really account for this organization. The social insects would clearly warrant a chapter to themselves, if the problems they pose had not so recently and so admirably been discussed by Hamilton (1964A, 1964B). Here I will give only a summary treatment of the subject. For further details the reader is directed to Hamilton's papers, which also anticipate a number of lines of reasoning developed elsewhere in this book.

The social insects typically live in colonies of one or both parents together with numerous offspring of different ages. The older offspring aid in the rearing of younger, or take over this duty entirely. The most remarkable development is the permanent sterility of a large proportion of the offspring (workers and other servile castes) in the more advanced societies. Such societies presumably evolve from family groups in which the parents raise their young in more or less permanent nests and attempt to manage broods of different ages at the same time. Under such conditions the older offspring may aid their younger brothers and sisters and make an important contribution to the success of the family, with even very limited self-sacrifice. A gene that augmented this behavior would be favorably selected because the aid provided would usually go to other individuals with the same gene. The duration of servitude to its parents' reproductive program might be an important factor. An extra day's assistance by an advanced termite nymph might make the difference between life and death for many of the younger and more helpless members of the colony. This time factor is necessary to account for the origin of the permanent sterility of the workers and soldiers of advanced societies. A gene that would extend servitude by one day would lower fitness in the juvenile, because it would provide it with an extra day in which it might be killed before reproducing, and because it would divert food energy and other resources that could serve its own reproduction. If and when such an individual matured, however, reproductive success might be substantially increased by the presence of the same gene in its progeny.

The accumulation of such genes, along with others that would make the rates of production of fertile offspring adaptively adjusted to seasonal or other ecological changes, might eventually result in permanent sterility for a large number of individuals. Such genes, in the aggregate, may make the probability of an individual's becoming a fertile adult only one in a hundred. Nevertheless they would continue to be favorably selected if they increased the success of the reproductives by more than a hundredfold. Such an evolutionary development would be based on the natural selection of alternative alleles in Mendelian populations, as is other adaptive evolution, but with a complication that need not ordinarily be recognized. There is an unfavorable selection within family groups that must be balanced by a favorable selection between groups. Alternatively one can recognize that an unfavorable character (likelihood of becoming a

SOCIAL ADAPTATIONS

worker) early in development might be balanced by a favorable character (high reproductive output) later on.

Unfortunately for this general picture, not all insect societies are "typical" in being headed by a single parental pair or fecundated queen. Some large colonies of ants may have more than one queen, and some primitively social bees may form colonies by the simultaneous contributions of several mature females. In some cases it may be misleading to call such females *queens*. The fact that several fertile females may occupy the same nest complex need not imply any real community organization. It may merely mean that the vicinity of other brood-rearing females is a favorable place to nest. Naturally each individual makes adaptive adjustments to the benefits and hazards that attend the close proximity of the other mature females, but each one has her own private chamber in the multiple nest and normally raises only her own young. Such colonies are no more socially organized than are those of colonial birds. When bee colonies from multiple founders do show more advanced organization it can be explained on the assumption that the founders are sisters. This is a factor that would favor the evolution of cooperative behavior on the basis of genic selection, and there is evidence (reviewed by Hamilton) that multiple queens or multiple founders are normally sisters.

The fact that highly elaborate insect societies exist is proof that, given the necessary evolutionary forces, genetically based animal sociality can achieve an advanced level of adaptive organization. That such societies seem to be based on an elaboration of family groups indicates their fundamental kinship with other

SOCIAL ADAPTATIONS

mechanisms by which organisms attempt to increase their currency of offspring. The apparent absence of comparable organization in any group of unrelated individuals is cogent evidence of the unimportance of biotic adaptation.

I believe that this is a valid interpretation of the more general features of insect societies. Nevertheless, there are some embarrassing details to be noted. Some of these problems exist on an abstract level. Most of the theoretical models on which discussions are based, including that outlined above, are based on the traditional theorems of population genetics, such as a Hardy-Weinberg distribution of genotype frequencies. Some of these assumptions are of dubious validity when applied to the social Hymenoptera. Matings, in one respect, are never at random, but always between a haploid genotype and a diploid genotype. Presumably all sperm from a given male carry the same genes, so that his offspring would show considerably greater genetic uniformity than would be found in normal Mendelian progenies. Sometimes, however, a queen mates with more than one male on her nuptial flight. No one has attempted an evaluation of the effects of such factors on the population genetics of the Hymenoptera. None of these complications applies to the termites, in which mates are selected for life and all individuals are diploid.

Other difficulties are at the observational level. The closeness of relationship between the individuals of a colony can sometimes be seriously questioned. Even though multiple queens are normally supposed to be sisters, they would inevitably be genetically different and produce genetically different offspring. Geno-

SOCIAL ADAPTATIONS

typic diversity within such sister-queen colonies would be significantly greater than in one-queen colonies, but presumably less than in the population as a whole. If it could be shown that there are thoroughly unified insect societies that normally contain several unrelated reproductives, they could only be explained as biotic adaptations resulting from effective group selection. The kinship of the reproductives would be a difficult proposition to prove one way or the other, but it is an extremely important point.

The model of selection based on the presumed close relationship of the members of an insect colony may also furnish an explanation for the somewhat analogous social structure of the California woodpecker (Ritter, 1938). These birds sometimes occur singly or in pairs, but more often they are found in more or less permanently localized "settlements" of a few to perhaps a dozen individuals. Such settlements have a particular tree in which the birds drill holes for the storage of acorns. All the individuals in the settlement contribute to the stocking of the storage tree as a community project and all make use of stored food as the need arises. When jays or squirrels try to rob the tree, one or more of the woodpeckers drive them away. Reproduction is also a community project, and more than two individuals are usually found in attendance at a nest.

There is very little evidence on the crucial question of genetic relationship. Some of Ritter's observations suggest that the birds may nest twice a year and that the young of an earlier brood may not leave when the new clutch of eggs is laid. Of the multiple birds that attend a single nest, some are manifestly immature and may represent older offspring helping with

SOCIAL ADAPTATIONS

the rearing of their younger brothers and sisters. I would predict that this would be the case, and that the societies of the California woodpecker, of the social insects, and of all other such organized groups, will be found to be based almost entirely on family relationship.

The relative genetic homogeneity of the offspring of a single pair of parents is probably of general importance as an evolutionary factor, even when it does not cause the development of elaborate social structures. It would undoubtedly be of advantage to a nestling bird to have the will and ability to push its nest mates out of the nest and appropriate all of its parents' solicitude for itself. But when this bird tried to raise young of its own, the genetic basis of its early success might have deleterious results. Competition among brothers and sisters in a nest seldom takes this extreme form, but a cowbird nestling usually evicts its normally unrelated competitors. Fisher (1930) pointed out a number of possible effects of this factor and showed that it had probably played an important role in the production of "distastefulness." A distasteful insect "educates" a predator on the unpalatability of its nearby brothers and sisters. Education on the unpalatability of the species as a whole is only an incidental effect. Fisher cited as evidence the marked tendency of distasteful species to hatch from large clusters of eggs and to live in sibling groups for some time.

Another illustrative example is found in the parasitic wasps (Salt, 1961). In some species, the mother lays several eggs on the same host. The resulting larvae seem to develop in perfect amity with each other. In other species, a mother lays only one egg

SOCIAL ADAPTATIONS

per host. When two such mothers each lay an egg on the same host, one larva will destroy its competitor and secure the host all to itself. In all these examples, the behavior of a normal group of brothers and sisters can be evaluated by comparison with a closely similar group in which genetic relationship is lacking. Even where we cannot make such a comparison, however, it is reasonable to assume that the relative genetic homogeneity of a Mendelian progeny is a factor that significantly softens the competitive interactions among brothers and sisters.

WHILE the extreme examples of cooperation and self-sacrifice in a group project are confined to genetically similar groups, it is nevertheless true that such behavior is sometimes observed among individuals that are not necessarily closely related. Some of the examples, such as primitive human social organization and some social interactions of a few other mammals, can be attributed to an evolutionary effect of the capacity of individuals to form personal friendships and grudges. This matter was explored in Chapter 4. The remaining possibilities are that the interactions represent biotic adaptation, that they represent a misplaced reproductive function, or that they are a statistical effect of individual adjustment. I favor misplaced reproductive function as the correct interpretation in specific individual interactions. The validity of this interpretation is supported to the extent that it can be shown (1) that reproductive functions do, in fact, get performed out of normal context, often to the detriment of all concerned, and (2) that when an animal actively assists an unrelated individual, it uses only those behavior patterns that are normally seen in a family setting. Benevolent behavior towards unrelated individuals should never be more intense, and should usually be less intense, than the same behavior towards offspring.

The first point is abundantly documented. The structures that relate to reproduction do not suddenly appear for the breeding season and then completely disappear. The gonads of an adult fish enlarge greatly as the spawning season approaches, but they are present from early in life and they do not disappear entirely between seasons. If perfect economy and efficiency were possible they would appear suddenly when needed and then vanish when the need had passed. The same can be said of accessory structures. Secondary sex dimorphism, for example, may become apparent well before the first breeding season, and may be detectable between seasons, even though its main development and only use are in reproduction. The same is manifestly true of behavior. Anyone who has observed the development of young animals, including children, can recall instances of abortive courtship or sexual conflict long before reproduction is actually possible. Nuptial singing, territoriality, and other behavior characteristic of birds in the springtime can sometimes be observed at lowered intensities at other seasons. In turtles, rudiments of sexual behavior can be observed almost immediately after hatching (Cagle, 1955).

Imperfections in timing are not the only examples of the looseness of control of reproductive functions. Homosexuality is an extremely common phenomenon in a wide variety of animals. It is of frequent occurrence among domesticated mammals, and is known

SOCIAL ADAPTATIONS

in wild ungulates (Koford, 1957) and wild monkeys (Altman, 1962). Freedman and Roe (1958, p. 468) maintained that "homosexual behavior seems to occur in all observed mammalian, primate, and human groups." It also occurs in finches and in sticklebacks (Morris, 1955). A male stickleback is said occasionally to show the "entire mating pattern of the female." Undoubtedly a large number of other examples could be listed. A related malfunction is courtship and attempted copulation with unreceptive pregnant females, as observed by Altman (1962).

Experienced canine enthusiasts will attest to the commonness of pseudo-pregnancy in bitches that have been through heat without being fertilized. Such animals may show all the behavioral symptoms of pregnancy, including the preparation of a littering retreat.

The production of hybrids is another example of the malfunctioning of the mechanisms of reproduction. Crosses between species are usually the result of abnormalities, such as those that prevail in captivity or are occasionally imposed on wild populations. Drastic alterations of habitats often result in the sudden appearance of hybrids in nature. This generalization is especially well documented for fishes, but a number of fish hybrids are known for which no abnormal circumstance can be demonstrated (Hubbs, 1955).

It would appear, therefore, that reproductive functions, perhaps to a greater extent than other adaptations, are characterized by a considerable degree of looseness in timing and of imperfection in execution. All of the examples listed above are clearly of evolutionary detriment to the individuals involved. At best, they wasted time and food energy that could have been constructively spent.

Reproductive functions, however, frequently involve the expenditure of time and energy by one individual in such a way that another (mate or offspring) is clearly benefited. Inevitably this sort of accessory reproductive behavior will also sometimes be found out of context. Certain rabbits and deer, when they take flight, raise their tails and display conspicuous markings. I interpret this as accessory to reproduction. The marking and the tail raising are designed partly to warn dependent young of the approach of danger, and mainly to attract the attention of a possible predator. Adults of such species have dependent young for a considerable proportion of their lives. In principle, it might be well to dispense with both the behavior and the markings whenever there are no young to be protected. In practice, it is not worth burdening the germ plasm with the information necessary to realize such an adjustment.

As a result, rates of predation on deer and rabbit populations, even out of the breeding season, are probably somewhat reduced by the warning signals that these animals display when they take flight. This circumstance means that the vicinity of conspecific individuals has value as protection against predators, and it undoubtedly contributes to selection pressures in favor of gregariousness in such species. These developments, however, involve no biotic adaptation. They merely represent individual adjustments to opportunities presented by their ecological environments.

Analogous warning signals are found in birds. The value of distraction displays as a protection for nes-

SOCIAL ADAPTATIONS

tlings is greatly enhanced by the white outer tail feathers of some species. These become conspicuous when the birds take flight, and the mechanics of flight make the display inevitable. That the bright outer tail feathers do not disappear between breeding seasons can be explained in the same way as the persistence of the white rump patches of mammals. The same argument applies to vocal warning mechanisms. I know of no examples of such warning devices in species that do not show well-developed parental care.

With certain types of population structure, those that often result in the continued close proximity of parents and offspring after the period of juvenile dependence, there might be weak selection pressures favoring the continuance of warning signals. Winter herds of deer, for instance, are often formed by the amalgamation of family groups. The rump-patch display by parents would continue to serve their own reproductive interests by warning the young, and the same behavior by the young would aid their sibs and parents. In large herds, however, such favorable selection would be largely offset by the fact that the displays would mainly aid genetic competitors.

Pairs of breeding birds are sometimes found to have one or more helpers, unmated individuals that assist the breeding pair in nest construction and other chores. As indicated above, this phenomenon may be of regular occurrence in the California woodpecker. It is observed sporadically in other species, and Skutch (1961) has provided a valuable review of this phenomenon. Helpers are known in a wide variety of taxonomic groups, and they are of both sexes and all ages, but ordinarily only a small propor-

SOCIAL ADAPTATIONS

tion of the breeding birds of a given area will have them. Helpers appear to be birds that have their own reproduction frustrated in some way so that they must find some other outlet for their parental instincts. Sometimes this frustration is only temporary, and a bird that functions briefly as a helper will succeed in finding a mate and in raising its own family. As expected, a large proportion of these temporarily or permanently frustrated birds are young adults having their initial breeding experience, and some may be manifestly immature. There are known cases of helpers assisting with the young of a different species.

The evolutionary explanation for this phenomenon should be obvious. The helpers are precisely the kinds of individuals that would be expected to show irrelevant reproductive behavior, such as homosexuality or misplaced parental care. The behavior that they do show includes only those elements, such as nest building and food gathering, that form a part of the normal reproductive behavior of the species. The helper phenomenon can be attributed to selection pressures for the maintenance of a certain pattern of parental behavior, with a less-than-perfect system of timing mechanisms for regulating this behavior.

THE EXAMPLES considered above all related to interactions between individuals, and the important consideration was to find a parsimonious explanation of why one individual would expend its own resources or endanger itself in an attempt to aid another. There remain a number of examples of individuals' acting, at their own expense, in a manner that benefits their conspecific neighbors in general, not specific individ-

SOCIAL ADAPTATIONS

uals. Such activity can take place only when the animals occur in unrelated groups larger than two. The important initial problem is why animals should exist in groups of several to many individuals.

It is my belief that two basic misconceptions have seriously hampered progress in the study of animals in groups. The first misconception is the assumption that when one demonstrates that a certain biological process produces a certain benefit, one has demonstrated *the* function, or at least *a* function of the process. This is a serious error. The demonstration of a benefit is neither necessary nor sufficient in the demonstration of function, although it may sometimes provide insight not otherwise obtainable. It is both necessary and sufficient to show that the process is designed to serve the function. A relevant example is provided by Allee (1931). He observed that a certain marine flatworm, normally found in aggregated groups, can be killed by placement in a hypotonic solution. The harmfulness of such a solution is reduced when large numbers of worms, not just one or a few, are exposed to it. The effect is caused by the liberation of an unknown substance from the worms, especially dead ones, into the water. The substance is not osmotically important in itself, but somehow protects the worms against hypotonicity. Allee saw great significance in this observation, and assumed that he had demonstrated that a beneficial chemical conditioning of the environment is a function of aggregation in these worms. The fallacy of such a conclusion should be especially clear when it relates to very artificial situations like placing large numbers of worms in a small volume of brackish water. The kind of evidence that would be acceptable would be the demonstration that social cohesion increased as the water became hypotonic or underwent some other chemically harmful change; that specific integumentary secretory machinery was activated by the deleterious change; that the substance secreted not only provided protection against hypotonicity, but was an extraordinarily effective substance for this protection. One or two more links in such a chain of circumstances would provide the necessary evidence of functional design and leave no doubt that protection from hypotonicity was a function of aggregation, and not merely an effect.

The second misconception is the assumption that to explain the functional aspects of groups, one must look for group functions. An analogy with human behavior will illustrate the nature of this fallacy. Suppose a visitor from Mars, unseen, observed the social behavior of a mob of panic-stricken people rushing from a burning theatre. If he was burdened with the misconception in question he would assume that the mob must show some sort of an adaptive organization for the benefit of the group as a whole. If he was sufficiently blinded by this assumption he might even miss the obvious conclusion that the observed behavior could result in total survival below what would have resulted from a wide variety of other conceivable types of behavior. He would be impressed by the fact that the group showed a rapid "response" to the stimulus of fire. It went rapidly from a widely dispersed distribution to the formation of dense aggregations that very effectively sealed off the exits.

Someone more conversant with human nature, however, would find the explanation not in a functioning of the group, but in the functioning of indi-

SOCIAL ADAPTATIONS

viduals. An individual finds himself in a theatre in which a dangerous fire has suddenly broken out. If he is sitting near an exit he may run for it immediately. If he is a bit farther away he sees others running for the exits and, knowing human nature, realizes that if he is to get out at all he must get out quickly; so he likewise runs for the door, and in so doing, intensifies the stimulus that will cause others to behave in the same way. This behavior is clearly adaptive from the standpoint of individual genetic survival, and the behavior of the mob is easily understood as the statistical summation of individual adaptation.

This is an extreme example of damage caused by the social consequences of adaptive behavior, but undoubtedly such effects do occur, and they may be fairly common in some species. There are numerous reports, at least at the anecdotal level, of the mass destruction of large ungulates when individuals in the van of a herd are pushed off cliffs by the press from the rear. Less spectacular examples of harm deriving from social grouping are probably of greater significance. I would imagine the most important damage from social behavior to be the spread of communicable disease.

The statistical summation of adaptive individual reactions, which I believe to underlie all group action, need not be harmful. On the contrary, it may often be beneficial, perhaps more often than not. An example of such a benefit would be the retention of warmth by close groups of mammals or birds in cold weather, but there is no more reason to assume that a herd is designed for the retention of warmth than to assume that it is designed for transmitting diseases.

SOCIAL ADAPTATIONS

The huddling behavior of a mouse in cold weather is designed to minimize its own heat loss, not that of the group. In seeking warmth from its neighbors it contributes heat to the group and thereby makes the collective warmth a stronger stimulus in evoking the same response from other individuals. The panic-stricken man in the theatre contributed to the panic stimulus in a similar fashion. Both man and mouse probably aid in the spread of disease. Thus the demonstration of effects, good or bad, proves nothing. To prove adaptation one must demonstrate a functional design.

I WILL discuss the origin and evolution of social grouping mainly with reference to the schooling of fishes, the example with which I am most familiar. A school in its typical development is a striking phenomenon. It is remarkable for its compactness and the way it moves about as a single unit. As with a well-prepared military drill, it is the regularity and precision of the group that excites interest, not the behavior of the individual participants. If one does look at individuals, one finds that they are all of the same species and of almost identical size, and that each fish always swims with very nearly the same speed and direction as its immediate neighbors. It is easy to lose interest in such individuals; if you have seen one you have seen them all. It is the school as a whole that persistently excites our curiosity.

I would maintain, however, that an understanding of schooling can be achieved only by counteracting this intuitive reaction. One should concentrate first on the individual and seek an understanding of the adaptive aspects of its behavior. If this inquiry is

SOCIAL ADAPTATIONS

successful, one can then ask how many of the phenomena of the school can be explained as simply the statistical summation of individual adaptations. Schooling can be expected to arise in any species that lives in habitats that are deficient in concealment from enemies, and in which there would be any tendency to form nonsocial groups. Such groups might form as a result of the localization of food, as at zooplankton concentrations where surface waters converge and descend. When a loose grouping is formed as a result of attraction to such a food source, the first individuals sighted and attacked by an approaching predator would be those on the periphery of the group. Under these circumstances, there would be selection in favor of individuals that had some genetic basis for being in the center of the group more often than their fellows. By placing itself inside the group, a fish would be putting others between itself and sources of danger from predators. It can provide itself with their protective presence in two ways, by actively placing itself in their midst and by summoning them by species-recognition marks. Thus schooling would be based on active behavior patterns and on passive exhibition of cues. Another likely development would be the evolution of the ability to recognize flight or distress among companions and to act defensively when such reactions are perceived. Gregariousness could be expected to increase until some factor, such as the depletion of food in the centers of schools, balances any further protection from increased intensity of schooling.

Breder (1959) reported a number of observations of predators attacking schools. An attack might be seemingly directed at the school but would usually

SOCIAL ADAPTATIONS

result in a single fish's being taken. Under these circumstances, the safest place for one of the prey was probably deep in the school. Only if the predator missed many of the more peripheral individuals would an interior fish be in danger, and it would have these other individuals' reactions to warn it of the attack. This being so, and having repeatedly been so in the history of the species, the individual reacts adaptively by avoiding the outside of the school. The same reaction, practiced by all the other individuals, increases the compactness of the school. Suppose an individual were to behave otherwise and swam away independently. It would immediately become the most conspicuous and most vulnerable of the prey. Any tendency to such nonconformist behavior would be unfavorably selected, and schooling would continue to be characteristic of the species. This conclusion follows regardless of the effect of schooling on predation on the group as a whole. It is quite likely that schooling by the prey sometimes facilitates predation. It may be easier for a predator to follow a large school than to search for isolated individuals. Many of the examples described by Breder (1959) suggest this. Some observations on predator behavior, notably those of Fink (1959), suggest that predators attempt to shepherd their prey so as to prevent schools from fragmenting. Certain behavior patterns and structures, such as those of the swordfish (Rich, 1947), are interpretable as adaptations designed to exploit the schooling behavior of prey. Sometimes a species may be effectively exploited by a predator only because it schools. This is true of predation by man. To be commercially exploitable, a species must

SOCIAL ADAPTATIONS

either be so large that it is economically feasible to pursue them one at a time, or it must school, so that a fisherman can pursue large numbers in a single operation. No small fish that does not school can support a profitable fishery. Bullis (1960) noted what might be a similar case. He observed a large shark feeding on a dense school of thread herring. The shark was said to bite off mouthfuls of the school in a manner that suggested a person eating an apple. The thread herrings would probably have been beneath notice as isolated individuals.

These observations hardly prove that a school is not adaptively organized, and they do not prove that some group benefit is not the function of a school, but they do show that schools may fail to show design for group survival in situations that might be expected to disclose such a design. It should also be said that there are probably times when schooling does reduce predation. Brock and Riffenburgh (1960) indicated one way in which such an effect might be produced.

The above account of the origin and function of schooling behavior requires only genic selection and recognizes only organic adaptation. It postulates that schooling behavior (the individual activity) is adaptive, but that a school (the statistical consequence) is not. A school and all its properties are explained as the statistical summation of the individual reactions. The extreme compactness of many schools is attributable to the reluctance of its members to be on the periphery. The unanimity of locomotory behavior results from each individual's trying to stay close to its fellows. The homogeneity of size and species results from a minimization of conspicuousness. Any

SOCIAL ADAPTATIONS

individual that looks different or behaves very differently from the others will stand out as an individual target. This explains why schooling may be intensified as a reaction to danger (Breder, 1959) or to the absence of cover (Williams, 1964). It explains why schooling behavior almost completely disappears at night in nearly all known cases.

There is one kind of property which, if it could be demonstrated to operate in any fish school, could not be explained by the theory that I have proposed. I refer to the possibility of warning signals, any reactions of alarmed fish that are not incidental to their own alarm reactions, and which stimulate alarm reactions in other fishes when perceived. Such signals could in no way benefit the senders. They would have to be considered biotic adaptations. To rule out the possibility of misplaced reproductive function, such warning signals would have to be demonstrated in a species in which they could not function primarily to warn the sender's own offspring. I know of no evidence for visual or sonic warning signals in fish schools. The possibility of chemical signals is discussed elsewhere (Williams, 1964: 377-378).

For all its striking regularities, a school is an outstanding illustration of unorganized social groups. In a herd of ungulates or a pack of wolves there is usually some apparent diversity of age. There may be some form of leadership, and perhaps a dominance-subordination hierarchy. Such individual distinctions are conspicuously absent in a typical school. Two separate schools can merge and mingle with no apparent resistance, and a single school can divide with equal ease. These are not the marks of adaptations produced by natural selection. The regularities of

SOCIAL ADAPTATIONS

schools lie in the statistics of redundancy, not in adaptive organization.

THERE are other groups that may approach fish schools in their homogeneity of membership and unanimity of activity: schools of squid, cetaceans, and sea snakes, and some nonbreeding flocks of birds. The discussion of the evolution and function of schooling behavior can be applied to these groups with slight modification. Rand (1954), however, produced evidence that protection from predators is not an important function of gregariousness in birds. He believed that facilitation of feeding was more important. He also produced some interesting examples of possible group damage resulting from gregariousness. I would suspect that large herds of ungulates may be functionally analogous with fish schools. Mammals, however, never more than partly abandon their roles in the drama of reproduction. A bull may have a seasonal cycle of behavior, but he retains a part of his belligerent nature at all seasons. The calf has a dependency that spans several seasons; and the cow may retain her attachment to the calf even after both join a large herd. No such complexities of family organization mar the perfect uniformity of the school. I would suggest that seasonal gregariousness of herbivorous mammals results from evolutionary forces analogous to those that cause the development of schooling behavior, but that the effect produced is seriously compromised by persistent family organization. Gregariousness in a wolf pack probably has an additional significance due to the wolfish tendency to attack large animals. A wolf can live on elk only when it attacks its prey in the company of other wolves

SOCIAL ADAPTATIONS

with similar dietary tendencies. I am not aware, however, of any evidence of functional organization of wolf packs.

The dominance-subordination hierarchy shown by wolves and a wide variety of vertebrates and arthropods is not a functional organization. It is the statistical consequence of a compromise made by each individual in its competition for food, mates, and other resources. Each compromise is adaptive, but not the statistical summation. Guhl and Allee (1944) took the opposite view. They showed that when a group of hens was first formed there was much fighting and other overt competition. This behavior decreased as the dominance hierarchy became established and the competition more ritualized. With this change there was an increase in average well-being, as measured by food consumption and egg laying, clearly the result of a decrease in each hen's expenditure of time and energy in overt competition. Guhl and Allee felt that the hierarchical organization itself must be responsible for the change. Wynne-Edwards (1962) also believes that dominance hierarchies are adaptive, not because they increase reproduction and food consumption, but because of evidence that they may decrease them (see pp. 234-246).

There are a few observations that suggest a functional organization in nonreproductive herds of mammals. When musk oxen are threatened by an enemy, the adult males may station themselves on the exposed side of the herd in what appears to be an attempt to defend the weaker members (Lydeckker, 1898; Clarke, 1954, p. 329; Hall and Kelson, 1959). This seems to be a functional division of labor and evidence for biotic adaptation, but there are other

SOCIAL ADAPTATIONS

possible explanations. It may be that the defending bulls are showing misplaced reproductive behavior. The herds are usually small, and if they contain a high proportion of the offspring of a defending bull, the behavior could be considered an organic adaptation directly relevant to reproduction. Or it could be that a purely statistical effect is at work. Each member of the herd would have a stimulus threshold that would determine whether it responded with counterthreat or with flight. Presumably this threshold is adaptively appropriate to the combative effectiveness of each class of individuals; a bull should be less easily intimidated than a calf. Accordingly there will be a range of intensity of threat stimuli that will evoke counterthreat by the adult males and flight or hiding by the less-well-armed members of the herd. Such statistical sorting does occur and can give a spurious appearance of biotic adaptation. When a herd flees in disorganized array, the adult males may flee with reduced intensity and lag behind the rest. This effect was observed in caribou by Murie (1935). Herds of bighorn sheep may segregate, with ewes and lambs staying close to the escape routes that lead to their high rocky havens, and the rams moving out on more level and more dangerous ground. That this is not a functional division of labor, with the rams there to protect the ewes and lambs, is apparent from the distance between the two groups. The rams may move, as a group, to points several miles from their more timid relatives (Blood, 1963). If the rams happened to be close to the ewes and lambs when an enemy appeared and if they reacted belligerently, they would give the appearance of the stronger attempting to defend the weaker individuals.

SOCIAL ADAPTATIONS

There are a large number of anecdotal reports of functional organization in groups of mammals. Many of these may be more a product of romantic imagination than of careful observation. Hall (1960) made a point of denying such organization in baboon troops, especially the supposition that individuals function as "sentinels" or perform other services for the group.

However, the possibility that such groups really do show a functional organization of some sort is sufficiently important to warrant careful attention by students of gregarious animals. These are likely groups in which to test the concept of biotic adaptation. Detailed and objective studies of wild populations, such as those by Altman, Hall, Lack, and Richdale, should provide important evidence on this point. That such studies have not yet furnished clear indication of the functional organization of large groups is already a matter of great significance.

CHAPTER 8

Other Supposedly Group-Related Adaptations

CHAPTERS 5, 6, 7, and most of the present chapter are devoted to interactions between individuals and to the role of the individual in the phenomena of populations. It is in these phenomena that we would be most likely to find evidence of biotic adaptation, and in these that its absence would be the most significant.

I have argued that organic adaptations are abundantly exemplified by interactions between individuals, but it must be conceded that such adaptation is most conspicuous in the physiology of single organisms. The principle of the precise adaptation of means to ends pervades every level of the physiologists' attention, from the regional to the molecular. The smallest protist is an endlessly intricate machine, with all parts contributing harmoniously to the ultimate goal of genetic survival. Any such distinct individual is expected to be genetically homogeneous. We do not expect to find genetically different individuals cooperating in a single somatic system. I would explain this in the same way I would explain the general absence of a functional social organization among genetically diverse individuals: only between-group selection could produce such organization, and this force is impotent in a world dominated by genic selection and random evolutionary processes.

OTHER SUPPOSEDLY GROUP-RELATED ADAPTATIONS

Genetic homogeneity of the soma is ensured by the mitotic cell divisions by which growth is mediated, and often by additional mechanisms that keep the system uniform. Immunological responses make it impossible for genetically different tissues of vertebrates to form functional associations. Burnet (1961, 1962) interpreted this immunological intolerance as originally a mechanism for the rejection of mutant somatic cells. Mutation is one source of genetic diversity. The fusion of adjacent growths would be another source in truly primitive organisms, which may have been very diffuse and poorly defined (see pp. 134-138). Primitive animals today have mechanisms that normally prevent the fusion of genetically different parts (Knight-Jones and Moyse, 1961). It seems to be the general rule that when two planktonic larvae of the same species of sponge or coelenterate settle so close to each other that they come in contact very early in their growth, they may fuse and enter into a functional somatic relationship. At more advanced stages, however, genetically different growths do not cooperate in producing the same somatic machinery, but seal themselves off from one another and maintain their individual identities. An increasing tendency to avoid fusion is apparent as one ascends the scale of histological specialization. It is difficult to get genetically different colonies of such animals as bryozoans and ascidians to fuse with each other, even in the earliest stages of their development.

Plants are much more tolerant of foreign tissue than are animals, as is readily demonstrated by grafting operations. Horticultural practice indicates that members of the same genus can often fuse into a single physiological system. A quince root system may

OTHER SUPPOSEDLY GROUP-RELATED ADAPTATIONS

supply water and nutrients to an apple trunk. In so doing it contributes to the genetic success of the apple, and as long as the relationship persists it will produce no offspring of its own. This, of course, is an historically unique situation not directly influenced by selection.

Fusions of genetically different plants do occur in nature, however. Bormann (1962) has shown that the root systems of adjacent pines and other trees may fuse. As the trees grow, a more vigorous individual may suppress the growth of others with which it is connected and incorporate their root systems into its own. In this way a mature tree may be genetically homogeneous above ground, but genetically diverse in the root system.

An important question is whether the relationship shortens or prolongs the reproductive life of the suppressed individual. Another is the possibility of mutual benefit, perhaps a firmer anchorage for the participants in a graft. Even more important is the question of whether the captured roots contribute actively to the genetically foreign shoot or are compelled to do so by the dominant individual. It may be that, in the initial stages of the relationship, it is uncertain which individuals will contribute and which will benefit. Each may be attempting to exploit the tissues of the others, and some must inevitably lose in such a contest. The hormonal, nutritional, and other relations between root-grafted trees are obviously matters of great significance to the present discussion.

The fusion of separate individuals into a single organized soma occurs regularly in the cellular slime molds. Burkholder (1952) has described the relevant phenomena in provocative terms. A population of

OTHER SUPPOSEDLY GROUP-RELATED ADAPTATIONS

amoeboid cells inhabits the soil, in which each individual lives independently, feeding and reproducing by fission. Then attraction stimuli emanate from scattered individuals in the population and cause a centripetal migration of the amoeboid cells. They come together in solid masses that then differentiate into a base, a stalk, and a terminal fruiting body in which spores are produced. The cells that form the stalk and base of the fruiting structure all die after sporulation. Cells that cooperated in the formation of these "somatic" structures sacrificed themselves so that certain terminally located cells could reproduce more effectively. Experiments have shown that morphological variants, which are presumably genetically different, can coalesce into a unified soma, and that the different types can be recovered from among the spores produced from the genetically mosaic individual (Filosa, 1962). The crucial questions for the interpretation of these facts relate to the frequency and degree of genetic variability among coalescing amoebae. Are fruiting structures regularly or only occasionally diverse genetically? If genetic diversity is the rule, are the cooperating amoebae normally from only two or three clones, or from many genetically different clones? If no more than two or three clones are normally represented in a fruiting structure, a stalk cell would, in assuming a somatic role, be favoring the reproduction of a group of cells that would usually contain a large proportion of individuals genetically identical to itself. Selection might favor the somatic sacrifice, although less effectively than in a genetically homogeneous system, and the behavior of the amoebae could still be interpreted as a purely organic adaptation. If the proportion of

OTHER SUPPOSEDLY GROUP-RELATED ADAPTATIONS

genetically identical cells is ordinarily small, biotic adaptation would be indicated. There may be some basis for doubting the generality of extensive genetic diversity in single somata. Sexual reproduction is rare or absent in these organisms (Bonner, 1958), and they are probably not widely dispersed. Bonner (personal communication) states that the wind is ineffective in distributing the spores and that dispersal depends largely on transport by water or animals in the soil. These factors may ensure that the amoebae coalescing in one soma would often all be of the same clone.

The frequency in nature of the formation of genetically heterogeneous somata and other related problems are matters of deep significance to biological theory in several respects, and warrant the close attention of students of higher plants, cellular slime molds, and other organisms in which the phenomenon may occur.

IT HAS been supposed by a number of biologists, first by Weismann (1892) and recently by Emerson (1960), that death from old age is a biotic adaptation. It supposedly benefits the population by removing the superannuated to make room for the young, and in helping to shorten generation time it may facilitate an evolutionary response to rapidly changing conditions. This theory has been severely and justifiably criticized by recent workers, especially by Comfort (1956), and only a brief summary of this criticism need be reproduced here: (1) Senescence is a generalized deterioration of all organs and merely causes an increasing probability of death with increasing age. There is no "death mechanism" with the

OTHER SUPPOSEDLY GROUP-RELATED ADAPTATIONS

appearance of an evolved adaptation. As Comfort stated, "Senescence has no function—it is the subversion of function." (2) For organisms in which maximum life span is known with any certainty, the age structure of wild populations indicates that death from old age almost never occurs. (3) There is no evidence that the length of generations, even if it is strongly influenced by senescence, is ever a limiting factor in evolutionary rate. On the contrary, among the animals at least, it would appear that long life and rapid evolution are often associated (e.g., *Elephas, Ursus, Homo*). (4) Even with the assumption of an evolutionary benefit from shortened generations, and certainly without it, it is difficult to imagine how senescence, as such, could be favored by selection. (5) There is a simple and plausible alternative theory of senescence as the result of organic adaptation.

This alternative theory is based on the relationship, first accurately stated by Medawar (1952), of selection pressures to age, or, more precisely, to the reproductive probability distribution. The extreme cases are easily appreciated; any variation in fitness that appears before sexual maturity will play a role in determining the entire breeding population. By contrast, variation in fitness that appears only at an age to which almost no individuals survive will play very little role in determining the extent to which different individuals reproduce. Selection may therefore favor genes that produce slight increases in fitness in youth, even if they also produce markedly deleterious effects later on. Elsewhere (Williams, 1957) I have attempted a general synthesis of the phenomena of senescence on this theoretical basis.

OTHER SUPPOSEDLY GROUP-RELATED ADAPTATIONS

Leopold (1961) has proposed some new arguments on the evolutionary and ecological significance of senescence, with special reference to the higher plants. His thesis is that senescence should be considered an aspect of adaptation in the plant life cycle, and I believe that most of his reasoning is valid. For example, he convincingly establishes that the senescence of leaves is adaptively related to their usefulness to the plant. When the leaves are first formed they are high up and well illuminated, but with additional growth of the shoot they are shaded by newer leaves and become less effective in photosynthesis. The plant then redeploys nutrients from the older leaves to the younger, where they can be put to better use, and this further impairs the usefulness of the older leaves. When a leaf's contribution is no longer worth the price of its maintenance, the plant mobilizes as much useful material as possible and then sloughs off the now useless part. Leopold's statement of his general conclusion is misleading, however, because he refers to this whole sequence as senescence. The deployment of materials in a way that maximizes their usefulness is clearly adaptive, and it is contrary to normal usage to call this process senescence. The deterioration and abscission of the older leaf is senescence, and is not adaptive. It represents a measurable loss of materials and energy, and is the price paid for an adaptive deployment of materials that is presumably worth the price.

The same can be said for the senescence of whole plant bodies. Leopold maintains that this whole-plant senescence enables the plant to be seasonally specialized in an adaptive way. This is true, but again, the seasonal death of an annual weed is not an adaptation in the sense of a positive goal to be achieved. It is merely the price paid for extremely rapid morphogenesis and the maximization of the survivorship of the overwintering seed stage. So senescence of any kind would be, in my interpretation, a loss to the individuals involved. Its presence in a life cycle has to be explained on the basis of its allowing a development that is presumably worth the cost. Senescence can be called adaptive only in the sense that some features of positive benefit would be impossible without it. If, for instance, a perennial plant could not develop an adaptation that would permit the abandonment of any of its leaves, it would be seriously handicapped and its likelihood of success greatly diminished. The possibility of resorting to physiological sacrifices for the attainment of biological goals is an important resource. In analyzing the evolution of such relationships, however, one must be careful to distinguish the goal from the sacrifice.

It has been assumed by some biologists (e.g., by Norman, 1949) that poisonous tissue is a biotic adaptation. The function of the toxin may be realized only after the death of the toxic individual, so that the individual is in no way benefited. The toxin would be designed to destroy the enemies of the species. Demonstration of such a design would necessitate the elimination of simple alternative possibilities, such as the toxic substance's being functionally a repellent and only incidentally toxic, or being entirely fortuitous.

The presence of toxins in the integument clearly suggests a repellent function, and this possibility is easily tested by observing predator reactions. Repel-

OTHER SUPPOSEDLY GROUP-RELATED ADAPTATIONS

lent functions are apparently common in amphibians and insects, and are sufficiently effective to have caused the evolution of mimics in the insects. Any toxicity of the repellent chemicals would be easily explained. To be repelled by a toxic compound is a generally expected animal adaptation. There would be a strong tendency for anything toxic to be distasteful, and a substance designed to be distasteful would often have a concomitant toxicity.

A different explanation must be sought for those organisms that have their toxins restricted to internal parts, or have toxins that are not distasteful. Various fishes and marine invertebrates may have internal poisons that are tasteless and produce their harmful effects only long after ingestion by a predator. The same is true of some toxic plants; certain deadly mushrooms are supposedly delicious.

The marine fishes and many invertebrates apparently owe their internal poisons to their diets. Halstead (1959) has studied the problem of toxic fishes in great detail and has found that the toxicity of even the most deadly species is a phenomenon that varies with season and geography, as we would expect if it is a dietary effect. The evidence implicates phytoplankton blooms, especially certain dinoflagellates. It would be absurd to maintain that these microscopic plants produce their toxins for the good of their populations. The toxin apparently has no effect on the herbivores that feed on the plants that produce it nor on the fishes that feed on the herbivores. The only possible explanation of this toxicity of marine dinoflagellates to terrestrial mammals is that the relationship is a purely fortuitous one. This must also be the

OTHER SUPPOSEDLY GROUP-RELATED ADAPTATIONS

explanation for many of the animal toxins found in the tissues of higher plants.

A venom is a special class of toxic substance. There is no doubt about its being designed to produce destructive effects. This design is apparent in pharmacological properties and in the existence of special devices for introducing the venom into the body of the organism to be damaged. Some venoms are offensive weapons found only in carnivorous animals. The venoms of the coelenterates and the snakes are used in a manner calculated to aid in the overpowering of prey. Other venoms are purely defensive adaptations, and to function as such they must produce pain very rapidly. Ultimate toxicity would be an incidental effect. The obvious way to produce pain, however, is to produce tissue damage, and any substance designed to produce rapid tissue damage at the point of entry is understandably likely to cause a generalized toxicity after it diffuses throughout the body.

As far as I know, all defensive venoms produce pain very quickly and are readily explained as repellents. There is no need to postulate that they are designed to kill. The venom of a nettle is a histamine solution (Burnet, 1962). A more appropriate substance for causing local pain in an animal could scarcely be imagined. The venom of the stingray (Halstead and Modglin, 1950), stonefish (Smith, 1951), weaverfish (Carlisle, 1962), and scorpions (numerous reports) all cause intense pain very rapidly. So do the defensive stings of the social insects, at least in proportion to their size. The stings of one group, the hornets, are of proverbial effectiveness as repellents. These painful observations all support the

OTHER SUPPOSEDLY GROUP-RELATED ADAPTATIONS

theory that the venoms are designed to produce an immediately unpleasant effect on the enemy and thereby protect the venomous individual or (in the social hymenoptera) the colony. The frequent death of a hymenopterous insect after stinging a mammal is probably not a part of the plan. It is an unfortunate accident resulting from the abundance of firmly anchored collagen fibers in the mammalian skin. These may catch in the barbs of the sting and cause the insect to damage itself fatally when it attempts to withdraw.

According to various romanticized discussions (which influence the thinking of biologists at the most formative stages of their development) the rattles on a rattlesnake's tail are designed to warn animals in general of the dangerous presence of a rattlesnake. Current herpetological opinion (Klauber, 1956; Schmidt and Inger, 1957, p. 273) seems to follow Darwin and hold that the rattle is indeed a warning, but is designed to benefit the snake by acting as "an advertising device by which larger animals are warned away, so that they neither tread upon the snake nor molest it" (*Origin*, Chap. 6). This mechanism is presumed to depend on the larger animals' learning that snakebites are harmful. It seems to me that there is a much better explanation. The rattles call the attention of an attacker to a harmless and unimportant part, away from the head where the important weapons are located. The outcome of a fight between a dog and a viper would depend very much on whether the dog initially seized the reptile by the head or by the tail.

There are many apparent examples of biotic adaptation to be found in boy-scout literature and other treatments of the folklore of natural history. One such example, that frogs call in order to aid their fellows in finding water, was recently presented as a new idea in the technical literature.

In dealing with the various possible kinds of biotic adaptation I have confined the discussion to the problem of whether the phenomena really operate in the manner envisioned, and whether they suggest any creative evolutionary forces besides the natural selection of alternative alleles. In the absence of objective or generally accepted criteria of population fitness (see pp. 101-107) it has seemed pointless to attempt an evaluation of whether a supposed adaptation really would contribute to the well-being of the group. The proponents of biotic adaptation have not been in agreement on the fundamental question of its demographic effect. Biotic adaptation is assumed by Wright (1945) to increase population size and so augment the rate of emigration, which is assumed to be the effective factor in competition between populations. Biotic adaptation is assumed by Brereton (1962), Snyder (1961), Wynne-Edwards (1962), and others to prevent population growth beyond some optimum level, with emigration being only one of many possible mechanisms. Others, e.g., Emerson (1960), see an additional function in the production of an age distribution that is some sort of optimum from the standpoint of evolutionary plasticity or some other factor. In many cases it is not at all apparent what sort of ultimate function is envisioned, as in many of the discussions of assistance provided to one individual by another.

There is one respect, however, in which there is general agreement. Always when biotic adaptation is postulated, its immediate or ultimate effect is the improvement of the situation from a traditional aesthetic point of view. It is assumed that: a population of vigorous individuals under heavy predator pressure is better adapted than one that is sickly and chronically starved; a population that divides its resources into stable individual territories is better adapted than one in which there is a chaotic scramble for resources; a population in which territory or social position is held by threat-display and recognition by neighbors is better adapted than one that maintains the social structure by frequent combat with effective weapons; a population with stable density, stable age distribution, etc., is better adapted than one in which such factors fluctuate widely; a population with limited fecundity and low juvenile mortality rates is better adapted than one with high fecundity and high juvenile mortality rates; a population in which the old and dominant individuals regularly yield to promising youths is better adapted than one dominated by a stable regime of fecund but slowly displaced oligarchs; populations in which individuals, such as worker bees, often jeopardize their own well-being for a larger cause are better adapted than those whose members consistently act only in their own immediate interests; those in which individuals normally live in peace or active cooperation and mutual aid are better adapted than populations in which open conflict is more in evidence; on the other hand, when active mutual destruction must take place, infanticide is preferable to the killing of peers. I submit that the only consistency found in such propositions is that they all conform to prevailing aesthetic concepts of what organisms ought to be like.

OTHER SUPPOSEDLY GROUP-RELATED ADAPTATIONS

Brereton (1962A) provided an interesting illustration of this aesthetic element in biological thought. He stated that the problem of the goal of biotic adaptation might be approached in a fanciful manner as follows (pp. 80-81): "... individuals in a population look around and say to themselves 'we are getting too numerous; mortality rate must be increased, or fecundity rate decreased, or both. If it is not, our standard of living will fall and we shall have to struggle with each other for survival.'" Brereton explicitly indicates that this discussion is metaphorical and eschews a belief in conscious community spirit as a factor in population ecology. Nevertheless, in his ensuing discussion of evolved mechanisms of population control, the adaptations envisioned all have in common that they raise standards of living in the anthropomorphic sense of an abundance of food and other resources for each individual. This is certainly a desirable aesthetic end in itself, but I see no reason for believing that it has an analogous evolutionary significance.

Brereton's discussion deals with the adaptive regulation of population size as the goal of evolved adaptations, which is certainly one of the more frequently assumed goals of biotic adaptation. Regulation here means negative feedback that contributes to stability. No one denies that population sizes in nature are frequently influenced by unique or random events that contribute nothing to the stabilization of population density. There are also some likely examples of positive feedback leading to population outbreak or irreversible declines to extinction. Nearly everyone, however, would admit the general tendency for natural populations to maintain more or less consistent

OTHER SUPPOSEDLY GROUP-RELATED ADAPTATIONS

levels of abundance for many generations, and most biologists would attribute this consistency to the continuous or at least occasional operation of stabilizing influences.

Such stabilization is most easily demonstrated in the very simplified ecology of experimental populations. One can, for example, set up an experimental container of water, inoculate it with a few protozoans, and keep it supplied with a constant increment of food per unit time. The numbers of the protozoan may then increase to a certain level determined by the rate at which food is supplied. Barring calamities the population will thereafter show only minor random or perhaps somewhat cyclic fluctuations about a long-term mean size. This equilibrium can be disturbed in either direction, by adding or removing individuals, and the population will rapidly return to its original level. Such a system clearly illustrates the phenomenon of a steady state maintained by negative feedback.

Many of the stable equilibria encountered by a biologist are the functions of evolved adaptations. The stability of mammalian body temperature over a wide range of environmental fluctuation is illustrative. This stability is achieved by the activities of a number of negative feedback mechanisms that must owe their existence and efficiency to selection for thermal stability.

We can likewise explore the causal factors that underlie the stability of an experimental population. As it increases in size, those factors that make for further increase are reduced, and those that tend to remove individuals become more effective. With increased population density, for example, rates of in-

OTHER SUPPOSEDLY GROUP-RELATED ADAPTATIONS

dividual development may be retarded by a number of factors: decreased food; toxification of the medium; interruption of feeding by competing individuals; etc. The same factors may interfere with the reproductive activities of the adults. Cannibalism or other active mutual destruction may increase rapidly with increased crowding. These negative influences intensify until the death rate is equal to the reproductive rate and the population stabilizes.

This sort of stabilization is often described in such terms as "the population adjusts itself to its food supply" or even "the population regulates its reproduction so as not to produce numbers in excess of what the environment can support." Such expressions imply that the density regulation is an evolved adaptation of the population as a whole, and that without such adaptations there would be no numerical stability.

These interpretations are utterly without justification. To maintain a certain number of protozoans or any other organism requires a characteristic expenditure of energy from the food supply. If the food is insufficient for such maintenance, the population must inevitably decline to a new level. This is a purely physical necessity. It is physically impossible for a population to exceed what its current environment is capable of supporting. The failure of a physical impossibility to occur is not something that we need attribute to evolved adaptations.

The regulation of such a simple experimental population is not, therefore, something that it accomplishes, but something that is sooner or later imposed upon it. An inquiry into the details of such regulation will, of course, deal extensively with adaptations as

OTHER SUPPOSEDLY GROUP-RELATED ADAPTATIONS

links in the chain of causes between food limitation and population limitation. All such adaptations, however, are designed to maximize the genetic survival of their individual possessors. Their effects on the population size will be a statistical by-product. If an animal gets just barely enough food to maintain its own tissues it can conceivably react in one of two ways. It can use part of its food supply to make gametes and thereby starve to death. Its death would then make more food available to other individuals, and this would tend to decrease their death rate. The other possibility is to adaptively adjust fecundity to food supply. On a bare subsistence diet the animal might stop reproducing entirely and just continue to subsist. Both alternatives, as a matter of physical necessity, result in a limited population. The adjustment of fecundity to food supply is adaptive from the individual standpoint. It permits the animal to survive, possibly to enjoy more prosperous times later on. From the population standpoint, however, this adaptation can do nothing either to promote or to inhibit the physically inevitable regulation of population size.

The introduction of such additional factors as predation may complicate the details but will not seriously disturb the outcome. Slobodkin (1959) has shown that experimental populations of *Daphnia* with a constant food supply maintain much the same population sizes under widely different conditions of predator pressure. As long as the predator (in this case, the experimenter) leaves enough of the prey to consume all the food, the food will be utilized to make new prey about as fast as they are removed by predation. Population biomass may be little affected

OTHER SUPPOSEDLY GROUP-RELATED ADAPTATIONS

by moderate predation, but its age structure and the conditions of existence for each individual may be greatly altered. Even with predation added to the system, the population ecology is extremely simple compared with what would normally be found in many natural communities. I would imagine, however, that some of the simpler and more homogeneous natural populations, such as those of the oceanic plankton, would be fairly well approximated by Slobodkin's experimental models.

Population sizes of most higher plants are difficult to study because of the problem of defining what would be meant by one individual, but the general problem is formally the same as it is for animal populations. Numbers and biomass are limited by a shortage of resources as a matter of physical necessity (Harper, 1960), and such limitation does not imply biotic adaptation.

In the examples discussed above the populations were limited, ultimately, by the limitation on the rate at which energy was fed into the system, but energy is only one of the necessary resources that may be in short supply. Bluebirds need holes in tree trunks, or appropriate substitutes, in order to nest. Unless some other resource is even more limited, the bluebird population will be determined by the abundance of suitable nesting holes. This limitation is inevitable, given the premise that holes of a certain description are a necessary resource. Likewise, given the condition that a certain kind of warbler is territorial and requires a minimum of an acre of woodland per pair, there will never be more breeding pairs of these warblers than there are acres of woodland. Although this kind of limitation is logically inescapable, it is

OTHER SUPPOSEDLY GROUP-RELATED ADAPTATIONS

not something that follows directly from physical principles. The resources in short supply are necessary only because certain evolutionary developments made them so. If a biologist were to devise a bird that would maintain the greatest possible abundance, he might give it leaf-eating habits and cellulose-digesting enzymes, small size, easily satisfied nesting habits, and a high degree of immunity to infectious diseases. He would not make it territorial. The acquisition of carnivorous diets, specialized nesting sites, territoriality, and other characteristics have necessarily resulted in reduced population densities, and some biologists have interpreted many of these developments, especially social interactions, as having the goal of population limitation. This tendency is quite apparent in the work of Allee (various works), Emerson (1960), Brereton (1962A and B), Snyder (1961), Wynne-Edwards (1962), and a number of other leading biologists.

The present discussion is especially concerned with the results of adaptation, the problem of whether regulation of population size is ever actually brought about by some attribute of the organisms rather than by operation of physical laws. Before considering the evidence, however, it is necessary to distinguish two aspects of population regulation. First, there is the *level* of regulation, the absolute value of the long-term mean of population size. Secondly, there is the *precision* of regulation, the closeness with which actual numbers conform to the long-term mean. One might conceivably demonstrate that territoriality limits a species to an average level of 100 pairs per square mile. This does not mean, however, that the existence of territoriality increases population stabil-

OTHER SUPPOSEDLY GROUP-RELATED ADAPTATIONS

ity. All that would be demonstrated is that the level of regulation is lower with territoriality than it would be without it. If territoriality did not limit the population, something else would limit it at a higher level, and this "something else" might regulate with greater precision or with less.

There are reasons for believing that territoriality and other forms of social-spacing mechanisms that operate independently of food and other resources can limit population densities (Lidicker, 1962). This means that regulation is at a lower level than would otherwise prevail, but until we can induce a whole species to change certain social characteristics and not change anything else, we will never know what the level or precision of regulation would be without the social spacing. All we know is that without this factor the level would be higher in some species. Lack (1954A) has shown that many territorial birds are very flexible in their acreage requirements, and that their densities are often well below what their minimum territory sizes would permit. It may be obvious that social spacing can limit populations, but that it actually does set the level of density in any particular case would be very hard to demonstrate. Even where such an effect could be demonstrated, the density regulation would not necessarily be the function of the social spacing. If the territoriality can be interpreted as an organic adaptation, the density regulation should be regarded as merely an incidental statistical by-product.

That territoriality may be of individual advantage is apparent from Lack's (1954A) demonstration that, in one species at least, high population densities (and concomitant small size of territories) result in reduced

OTHER SUPPOSEDLY GROUP-RELATED ADAPTATIONS

average success in reproduction. It is apparently advantageous to nest in areas of low density, and this factor may partly compensate for suboptimal conditions of other kinds. Two coordinated adaptations, the territorial display and the reaction to such displays in other individuals, contribute to the goal of nesting in areas of low density. The territorial display is a ritualized advertisement of determined occupancy of a certain area. When a bird is seeking a nesting site and continually encounters territorial displays by other individuals, it is obviously searching in an area of high population density. The adaptive reaction is to move on to less crowded regions. When an otherwise desirable site is located and no competitors advertise their prior claims, the bird may occupy the area as its own territory. Having done so, it becomes important that the advantage of low population density be maintained. To this end the bird commences a territorial display of its own, and thereby discourages potential competitors from utilizing any part of its occupied territory. Once a bird has occupied an area for a while, it acquires a vested interest that makes that particular territory of special value. During its stay a bird may learn the important aspects of the geography of its territory, the specific locations of refuges, of water, of food sources, and of alternative nesting niches. It may invest time and effort in nest construction. Its motivation for continued use of the territory should be higher than the motivation of a newcomer to appropriate it. Hence the common observation that territory holders almost always win in conflicts with intruders. It has been supposed that the effectiveness of territorial displays must always depend on the possibility of resorting to effective

OTHER SUPPOSEDLY GROUP-RELATED ADAPTATIONS

weapons if the display should fail to intimidate a rival. This is undoubtedly an important factor in maintaining the effectiveness of the ritualized and less damaging forms of conflict. Even small song birds may be injured or killed in the fights that sometimes occur (Smith, 1958, p. 238). It is nevertheless possible that even without the prospect of damaging combat, the desirability of nesting in uncrowded areas may normally ensure the retreat of intruders from ritually defended territories.

The precise way in which population density and territory size affect nesting success is not really clear. Lack (1954A, Chap. 22) questioned the traditional view that large territories are desirable because they hold more food for nestlings. He showed that success was influenced by territory size even when food seemed to be abundant in even the smallest territories. Tinbergen (1957) defended the traditional view by pointing out that the apparent long-term abundance of food may not be as important as its minimum abundance in relation to demand. The supply may be lowest and the demand greatest on a cold rainy day, and even one such day may have serious consequences for the success of a brood. Thus the value of a territory may lie more in its provision for a safe minimum than in its mean resources. Territoriality in animals that do not forage in their holdings for food for their young, for instance various fishes, pinnipeds, and marine birds, might be interpretable as the defense of some other kind of resource, such as a nesting niche or a harem. Freedom from competition in the courtship ritual is another possible benefit from defending a large territory.

I maintained in Chapter 6 that a well-adapted organism will assume the burdens of parenthood only when there is a favorable ratio between the probable risk and the prospects for success. Such things as physiological well-being, the availability of the proper kind of nesting site, and the presence of another individual showing the appropriate social releasers would be indications that prospects for reproduction are favorable. Disease, malnutrition, and shortages of needed resources would mean that efforts at reproduction should be minimized or postponed. Another indication of low prospects for success in relation to cost would be an unfavorable social environment. If an animal is continuously exposed to aggressive competition it may be adaptive, even if the animal is currently healthy and well fed, to curtail or delay its efforts to reproduce. Frequent contact with competitors would often be a reliable sign that to reproduce at all would require great effort and hazard to the somatic investment, and that any resulting offspring would be exposed to shortages and perhaps accidental death from the competitive interactions of the adults. It may be adaptive to act on the assumption that reproduction is not worth the effort in crowded situations. The expected reaction would be to postpone reproduction and search for a less crowded area. There is no reason to suppose, as Wynne-Edwards and others have done, that such restraints on reproduction are designed to prevent the population from overtaxing its resources. They are adequately explained as adaptations whereby an individual can adjust to the probable outcome of trying to reproduce.

All physiological mechanisms can fail if the stresses are great enough, and it is understandable that pro-

OTHER SUPPOSEDLY GROUP-RELATED ADAPTATIONS

longed exposure to an acutely unfavorable social environment would result in pathological symptoms. When oppressive social conditions are prolonged and escape impossible, either because crowding is continuous over a large geographical area, or because it is artificially maintained, psychological damage may result. Such results are seen in artificially crowded populations in captivity and in some very dense natural populations, such as those of lemmings during their occasional migratory outbreaks. These famous migrations are interpreted by Allee *et al.* (1949) as a biotic adaptation for stabilizing the population. An alternative view is implied above: the migrants are psychotically disturbed individuals and show, not a biotic adaptation, but a psychic malfunctioning induced by crowding. A number of facts support this interpretation. The migrants are mostly males; the removal of females would be a more effective way of controlling the population. The migrants are physically abnormal, in that they show histologically demonstrable endocrine disorders, which parallel those seen in rodents kept under very crowded conditions in captivity. The migrants do not really show indications of the suicidal intent attributed to them by popular writers. When they reach water they may detour, or they may attempt to swim across. They may drown in so doing, but they may also find safety on another shore. When favorable environments are reached, they may stop migrating and settle down. See Elton (1942), Frank (1957), and Thompson (1955) for detailed discussions of this phenomenon.

The most elaborate development of the theory that population density is adaptively limited by social be-

OTHER SUPPOSEDLY GROUP-RELATED ADAPTATIONS

havior is that of Wynne-Edwards (1962). Territoriality and related phenomena figure heavily in his discussion. Many of his arguments proceed by postulating that a certain event, such as mortality, occurs with a certain relative frequency in a population, that this frequency indicates a certain need for a regulatory function, and that this population need is met by a certain reproductive restraint on the part of its individual members. For every such theory, a perfectly parallel theory can be developed on the postulate that the event in question has a certain probability per unit time for the individual, and that such things as life expectancy at a certain age influence the kind of reproductive behavior that would or would not be adaptive for the maximization of individual reproduction. In earlier discussions of reproductive physiology I maintained, for example, that some species, such as eagles, have low intensities of reproductive effort because of a low probability of death from one breeding season to the next. Wynne-Edwards, by contrast, would explain the same observation by saying that the low population mortality rate requires low reproductive rates in order to prevent crowding. Considerations of parsimony would seem to favor my explanation over Wynne-Edwards', and there are some additional reasons for rejecting Wynne-Edwards' approach. He maintains, for example, that the function of sexual conflict is to limit the number of individuals that reproduce in a given breeding season, but for most of the species he cites, it is only the number of breeding males that would be strongly influenced. The number of participating females and the number of young produced would be little affected. That the sex that produces a great

OTHER SUPPOSEDLY GROUP-RELATED ADAPTATIONS

excess of gametes should be the one that plays the primary role in sexual conflict, territoriality, and courtship is a serious objection to the theory that these behavior patterns represent adaptations for the regulation of population size. As was indicated in Chapter 6, the male appropriation of these functions is readily explained as organic adaptation. For additional arguments against Wynne-Edwards' theory see Braestrup (1963) and Amadon (1964).

MOST of the proposed examples of biotic adaptation have involved interactions among members of one species or between a species, as a group, and some environmental factor of supposedly common interest, such as a predator. Some of the proposed examples, however, transcend the species level in organization. There are many examples of mutually beneficial relations between species, such as the algal-fungal coactions in lichens, the mutualism between termites and their intestinal biotas, many specific insect-angiosperm dependencies, and many of the obligate commensalisms of organisms, such as reef-corals, that produce large, structurally defined modifications of the environment. These phenomena have been interpreted as indicating that a species-complex is a unit of selection and adaptive evolution. This is certainly true in a sense. Neither a termite nor its intestinal symbionts can become extinct without the other sharing its fate. Likewise the evolution of each would have been very different had the other not been there. The important question, however, is whether the selection of alternative alleles can simply and adequately explain the origin and maintenance of such relationships.

OTHER SUPPOSEDLY GROUP-RELATED ADAPTATIONS

I believe that such an explanation is possible and plausible in every instance. We can expect cooperative mutualistic mechanisms to arise between any two species in which each constitutes, for the other, an important source of some aid to survival. Such a chance relation between species A and B would mean that close association with the other and aid provided to the other would be adaptive for both. "Helping A" would be adaptive for B as well as A. To a large extent, then, A and B would be expected to join and cooperate in the same strategy. The really good examples of mutualism are relatively rare, and it must be that these necessary preconditions seldom arise. If A benefits B but is harmed in providing the benefit, the two species will act at cross-purposes; B will continually try to establish a relationship that A will as consistently strive to avoid. The interaction between hound and hare or between flea and hound are clear examples of this extremely prevalent sort of relationship.

Ecosystems and, perhaps, the whole biota of the Earth have been considered adapted units (Allee, 1940). Animals and other heterotrophs are entirely dependent on autotrophic plants for their food, whether they consume plant parts directly or by eating herbivores. The autotrophs, on the other hand, benefit from the rapid return of raw materials by the heterotrophs. A balance and mutual dependence can certainly be recognized in this picture, but certainly not adaptation or functional organization. There is nothing in a (wild) carrot or other taproot to suggest that it was designed to be eaten by rabbits or by anything else. Its structure forces the conclusion that it is designed for food storage for the plant of which it

OTHER SUPPOSEDLY GROUP-RELATED ADAPTATIONS

is a part. Similarly the structure and behavior of a rabbit are more readily interpreted as means for escaping from predators than for supplying them with food. An ecosystem, as a machine, is highly inefficient for just this reason, the impediments raised by each trophic level to the passage of energy to the next higher level. It would seem absurd to belabor such an argument, but this is the critical evidence on the validity of the organization of the community as a concept in any way analogous to the organization of an organism.

Most discussions of these points are couched in language that makes it difficult to tell whether an author is recognizing evolved adaptations or fortuitous effects. It is certainly fortunate and necessary from our point of view that photosynthesis was evolved. It is also obvious that, if this mechanism had not appeared, the Earth would now be nearly barren of life. One can recognize these facts, however, without suggesting that any sort of advantage in supporting a large body of heterotrophs was active in any way in the perfection of photosynthesis. Here, as in all other problems of adaptation, it is important to distinguish functions from fortuitous benefits.

There are some noteworthy exceptions to the interpretational difficulties mentioned above. Some authors have been quite clear in their suggestions that community organization shows evolved adaptations superimposed on those of the component populations and individuals. A representative recent example is provided by a paper by Dunbar (1960). He assumes initially that large and erratic fluctuations in the size of populations, such as are often found in Arctic communities, are indications of poor adaptation. Con-

OTHER SUPPOSEDLY GROUP-RELATED ADAPTATIONS

versely, a high degree of stability in population size, as is found in the tropics, indicates a well-adapted community. He then argues that Arctic communities are new and have not had time to evolve stabilizing mechanisms, such as presumably prevail in equatorial regions.

Dunbar's analysis of the factors of stabilization indicates that variability of the physical environment is important. Such variations, not only from season to season but also from year to year, are much greater in high than in low latitudes. Any dependent fluctuations in population sizes would therefore be greater in Arctic than in Austral regions. Another factor is the paucity of the far northern biota compared with that of warmer regions. It means that the northern biota is characterized by a larger incidence of specific dependencies between species. An epidemic among snowshoe hares may have dramatic effects on the population of Arctic foxes, but the sharp reduction of a tropical herbivore population need not seriously affect any carnivore, because a tropical carnivore would have many possible species of prey. The prey, in turn, would probably exploit many species of plants and would be less likely to show great fluctuations as a result of changes in plant populations.

The explanation to this point involves nothing that suggests adaptations for stabilizing the community, and I fail to understand why it is not considered adequate. However, Dunbar goes on to suggest that species evolve lowered reproductive rates because this acts to stabilize community composition. That a general reduction in fecundity would, indeed, have such an effect might be debated, but any such curtailment, beyond what would be expected of adaptations de-

OTHER SUPPOSEDLY GROUP-RELATED ADAPTATIONS

signed to maximize each individual's currency of offspring would have to be explained by something other than genic selection. All of Dunbar's examples, however, such as the production of fewer (but larger) eggs, and the restriction of reproduction to seasons that optimize the food supply of the young, are clearly factors of organic adaptation to the special conditions of the Arctic. They are attributable to the action of genic selection on reproductive physiology and behavior, as indicated in Chapter 6. To attribute to such adaptations the additional function of community stabilization is gratuitous and unnecessary.

CHAPTER 9

The Scientific Study of Adaptation

THE PRECEDING discussions have portrayed a certain view of natural selection and advocated this view as the only acceptable theory of the genesis of adaptation. Natural selection arises from a reproductive competition among the individuals, and ultimately among the genes, in a Mendelian population. A gene is selected on one basis only, its average effectiveness in producing individuals able to maximize the gene's representation in future generations. The actual events in this process are endlessly complex, and the resulting adaptations exceedingly diverse, but the essential features are everywhere the same.

The significance of a Mendelian population is that it is a major part of the environment in which selection takes place. The population gene pool is the genetic environment of every gene. For each individual the population may be an important ecological factor in a variety of ways. It may provide some important resources, competition for other resources, and a social structure that favors the possession of specific social adaptations. The population parameters assign, to each individual, its age-related probability distributions of death, of reproduction, of specific kinds of stresses, of sex-ratios among social contacts, and of measures of spatial and ecological vagility. Such aspects of the demographic environment are factors to which organisms are precisely adapted, but this adaptation is often neglected in evolutionary discussions because of the tendency to think of a population as something adapted, rather than an environment to be adapted to.

A neontological species is a group of one or more populations that have irrevocably separated from other populations as a result of the development of intrinsic barriers to genetic recombination. The species is therefore a key taxonomic and evolutionary concept but has no special significance for the study of adaptation. It is not an adapted unit and there are no mechanisms that function for the survival of the species. The only adaptations that clearly exist express themselves in genetically defined individuals and have only one ultimate goal, the maximal perpetuation of the genes responsible for the visible adaptive mechanisms, a goal equated to Hamilton's (1964A) "inclusive fitness." The significance of an individual is equal to the extent to which it realizes this goal. In other words, its significance lies entirely in its contribution to one aspect of the vital statistics of the population.

Acceptance of these conclusions means that some widely used concepts are invalid and must be abandoned. The question inevitably arises as to how such an abundance of misinterpretation has arisen. I believe that the major factor is that biologists have no logically sound and generally accepted set of principles and procedures for answering the question: "What is its function?" In practice this question is answered on the basis of a variety of criteria, some of which are of value, but their use is largely dictated by taste and intuition and their value obscured by terminological inconsistencies.

THE SCIENTIFIC STUDY OF ADAPTATION

A frequent and often useful procedure is to rely on analogies between biological adaptations and human artifice. We can thus understand a structure such as a mammalian oviduct as a mechanism for conveying an ovum and early embryo to the uterus. Other analogies help us to understand the uterus as designed for the protection and nourishment of the embryo and foetus. We may similarly recognize the whole complex of reproductive machinery of both sexes as having the goal of producing viable offspring. But why are offspring produced? Is it, as is often stated, for the perpetuation of the species? Or are they produced, as I have maintained, to maximize the representation of the parental genes in the next and subsequent generations? There can be no more important question asked about mammalian reproduction, but there is no established procedure for answering it.

Analogies between biological phenomena and human affairs can also be of value at the level of groups of individuals, but there is no simple and reliable guide to tell us where to stop. Certainly there are interesting parallels between man and animal in family organization, especially between extended multigeneration human families and the colonies of social insects. There may be interesting parallels at even higher levels. A species has a continuity beyond the lives of its individual members; so has a nation of men. A species maintains itself by the activities of its members despite destructive external influences; so does a nation. But does a species have anything at all akin to a spirit of nationalism? A New Frontier? A Five-Year Plan? Does a species have a collective will to avoid extinction or anything at all similar to such a collective interest? No modern biologist has ex-

THE SCIENTIFIC STUDY OF ADAPTATION

plicitly proposed that such factors are operative in the history of a species, but I believe that biologists are unconsciously influenced by such thinking, and that this is true of some distinguished and capable scholars.

I see no other way of interpreting Cott's (1954) statement that conspicuous colors and behavior characterize the "less valuable" members of a species; or Amadon's (1959) statement that males of polygynous birds can live dangerously because "males need not be as numerous as females." Cott's reference to the "value" of an individual is apparently not concerned with its value to itself. Amadon obviously did not refer to how numerous a particular male genotype is. Nor are these uses of the concept of value and need being discussed from the standpoint of man's economic or aesthetic interests. They are being discussed from a species point of view as if there were some kind of collective interest that must be served.

Uncritical analogy with self-conscious human organizations is probably not the whole explanation for the humanization of groups of organisms. There may also be a desire, unconscious in many and expressed by a few, to find not only an order in Nature but a moral order. In human behavior a sacrifice of self-interest and devotion to a suprapersonal cause is considered praiseworthy. If some other organisms also showed concern for group welfare and were not entirely self-seeking, these organisms, and Nature in general, would be more ethically acceptable. In most theological systems it is necessary that the creator be benevolent and that this benevolence show in his creation. If Nature is found to be malicious or morally indifferent, the creator is presumably malicious or indifferent. For many, peace of mind might be difficult with the acceptance of either of these conclusions, but this is hardly a basis for making decisions in biology. There is a rather steady production of books and essays that attempt to show that Nature is, in the long run and on the average, benevolent and acceptable to some unquestionable ethical and moral point of view. By implication, she must be an appropriate guide for devising ethical systems and for judging human behavior. In some cases it would appear that "love thy neighbor" must stand or fall according to whether mutualism or parasitism is the more prevalent phenomenon. Attempts to demonstrate the benevolence of Nature often take the form of name changing. The killing of deer by mountain lions meant "nature red in tooth and claw" to a generation of "social Darwinists." To a more recent generation it has become Nature's kindness in preventing deer from becoming so numerous that they die of starvation or disease. To Darwin himself there was a poorly defined "grandeur" in such processes. The simple facts are that both predation and starvation are painful prospects for deer, and that the lion's lot is no more enviable. Perhaps biology would have been able to mature more rapidly in a culture not dominated by Judeo-Christian theology and the Romantic tradition. It might have been well served by the First Holy Truth from the Sermon at Benares: "Birth is painful, old age is painful, sickness is painful, death is painful . . ." (attributed to the Buddha by Burtt, 1955).

Of the terminological problems, *organized* and *organization* are especially troublesome. When a biologist says that a system is organized, he should mean organized for genetic survival or for some subordinate goal that ultimately contributes to successful reproduction. For a given organism we can usually specify a particular sort of approach to the problem of survival. An *Ascaris*, for example, is organized for survival in the intestines of horses. It has adaptive mechanisms that minimize the disadvantages of such a way of life and that help it to exploit the advantages of intestinal parasitism. Each part of the animal is organized for some function tributary to the ultimate goal of the survival of its own genes.

Organization attributed to groups should be subject to the same considerations. A family group of nesting birds is clearly organized for genetic survival. One critical function, that of rapid, efficient, and precise morphogenesis, is performed by the nestlings. The adults provide the germ cells for the production of the young and then supply the heat necessary for successful development, and after that they may supply food. There may be a division of labor between the parents and this too is designed to promote genetic survival for each of the parents. The most elaborate nonhuman social organizations—in numbers of subordinate individuals and in complexity of division of labor—are those of the social insects. These animal societies show parallels to ships' crews, ball teams, and other human organizations, which have the advantage of rational planning and cultural tradition.

This basically functional nature of biological organization may seem obvious, but it is possible to overlook it, because of the shifting meanings of the term "organization." Even the most chaotically disorganized system may have a precise statistical organization. The statistics of chaos or randomness is a basic standard for statistics in general. A parameter of any collection of entities may have a precisely specifiable arithmetic mean and other measures of central tendency, a precisely determined variance, skewness, kurtosis, and so on. The precision with which such population parameters may be maintained is not necessarily an indication of functional precision, it merely indicates a statistical constancy. The fact that a collection of organisms, whether it be a Mendelian population of insects, a herd of buffalo, or a pound of peanuts, has a precise mean size, weight, mutation rate, age distribution, etc., means that these parameters can be statistically organized. It does not mean that they are functionally organized.

Examples of groups of organisms that are functionally organized were mentioned above, but, as I argued in Chapter 7, not all groups have such organization. Each type of group must be examined separately to determine whether its characteristics make functional sense. Such an examination of a family-group of birds or of the bees in a hive unmistakably favors the recognition of a functional organization. An examination of other groups, such as a swarm of moths around a lamp or a mass of mussels on a piling, force no such conclusion. It is certainly possible that some groups have a functional organization that is too subtle to be appreciated by the conceptual and technical equipment of the observer. Considerations of parsimony, however, demand that we not recognize a functional organization unless we have definite evidence for it. We should not invoke biological principles where statistics suffices. In Chapter 8, I argued

THE SCIENTIFIC STUDY OF ADAPTATION

that most of the examples of gregariousness and other group phenomena can be explained as the statistical summation of individual adaptation and require no recognition of a functional organization of the group. One aim of this book is to convince the reader that an understanding of the general nature of adaptation is important and that its study requires a more rigorously disciplined treatment than it ordinarily receives. Indeed, I believe that it is important enough to warrant a special branch of biology for its investigation, and in the remainder of this chapter I will suggest ideas that might aid in the development of such a special field of inquiry.

The most urgent requisite for the success of any science is that it have a name. Pittendrigh (1958) suggested that the explicit recognition of the functional organization of living systems be called *teleonomy*. This term would connote a formal relationship to Aristotelian teleology, with the important difference that teleonomy implies the material principle of natural selection in place of the Aristotelian final cause. I suggest that Pittendrigh's term be used to designate the study of adaptation.

Teleonomy would not be a branch of the study of evolution. Its first concern with a biological phenomenon would be to answer the question: "What is its function?" An initial assumption might be, in Pittendrigh's words, that "some feature of the organism —morphological, physiological, or behavioral . . . serves some proximate end (food getting, escape, etc.) that the observer believes he can discern fully by direct observation and without reference to the history of the organism." I know of no better illustration of this process of recognizing functional design from careful observation than that provided by Paley (1836) in an answer to the suggestion that the eye just happened to be suitable for vision:

> ... that it should have consisted [by chance], first, of a series of transparent lenses (very different, by the by, even in their substance, from the opaque materials of which the rest of the body is, in general at least, composed; and with which the whole of its surface, this single portion of it excepted, is covered) secondly of a black cloth or canvass (the only membrane of the body which is black) spread out behind these lenses so as to receive the image formed by pencils of light transmitted through them; and placed at the precise geometrical distance at which, and at which alone, a distinct image could be formed, namely, at the concourse of the refracted rays: thirdly, of a large nerve communicating between this membrane and the brain . . . is too absurd to be made more so by any augmentation.

Any such plausible demonstration of design in relation to a goal would provide the answer to the teleonomist's prime question. His next task would be to explain why the mechanism in question is maintained as a normal characteristic of the species and not allowed to degenerate. His initial attempt would be to explain the mechanism as the inevitable consequence of the natural selection of alternative alleles in an environment described in relevant genetic, somatic, and ecological (including social and demographic) terms. As I have indicated many times, I believe that this attempt can almost always be successful, but if the attempt fails, a teleonomist may explore other

THE SCIENTIFIC STUDY OF ADAPTATION

possibilities, such as group selection or even mystical causes if he is so inclined.

Pittendrigh's assumption that the proximate end of an adaptation is something that a biologist "can discern fully by direct observation" may be too optimistic. The lateral lines of fishes and the singing of birds were directly observed for centuries without their immediate functions being fully discerned. In other cases a proximate end was discerned, but erroneously, as in the recognition of wings in the enlarged pectoral fins of a "flying" gurnard. And what is the purpose of increased melanin production in human skin exposed to sunlight? The immediately obvious answer turns out, on detailed investigation, to be at best only partly correct (Blum, 1961). More remote goals are also of teleonomic importance, but may be even more difficult to determine. Bird songs function as aids in the maintenance of territory, but what is the function of territory? A variety of alternative answers may be found in the recent literature.

How, ultimately, does one ascertain the function of a biological mechanism? In this book I have assumed, as is customary, that functional design is something that can be intuitively comprehended by an investigator and convincingly communicated to others. Although this may often be true, I suspect that progress in teleonomy will soon demand a standardization of criteria for demonstrating adaptation, and a formal terminology for its description. When this need arises, teleonomists will find many useful suggestions in the work of Sommerhoff (1950), who proposed just such a system of criteria and symbolism for dealing with adaptation. I was tempted to use Sommerhoff's system in this work, but decided that its general unfa-

THE SCIENTIFIC STUDY OF ADAPTATION

miliarity would be too great a disadvantage. Also, I believe that the system as proposed is suitable only for facultative individual responses, and will need revision for use with obligate adaptations.

Perhaps the main reason why biologists have not adopted a formal system for determining functional relationships is that many of the problems are so readily solved intuitively. We do not need weighty abstractions to help us decide that the eye is a visual mechanism. Also there are many helpful parallels between natural and artificial mechanisms, and it is so convenient as to be inevitable that parallel terminology be used. The close analogy between the lens of a camera and the lens of an eye make the term *lens* appropriate for both. From the teleonomist's point of view it is most important that these terminological transfers be made only when there is a real functional analogy between what man's reasoning (and trial and error) can produce and what natural selection can produce. One should never imply that an effect is a function unless he can show that it is produced by design and not by happenstance. The mere fact of the effect's being beneficial from one or another point of view should not be taken as evidence of adaptation. Under these rules it is entirely acceptable to conclude that a turtle leaves the sea to lay eggs, but not that a lemming enters it to commit suicide.

Parsimony demands that an effect be called a function only when chance can be ruled out as a possible explanation. In an individual organism an effect should be assumed to be the result of physical laws only, or perhaps the fortuitous effect of some unrelated adaptation, unless there is clear evidence that it is produced by mechanisms designed to produce it.

THE SCIENTIFIC STUDY OF ADAPTATION

In groups of organisms an effect should be ascribed completely, if possible, as the fortuitous summation of individual activities, unless there is evidence of coordinated teamwork for producing the effect, or mechanisms for producing group benefit by individual self-sacrifice. One should postulate adaptation at no higher a level than is necessitated by the facts.

When an organic adaptation is recognized, its explanation by genic selection can take one of several forms. It might be a fixed feature of the organism, or it might be facultative. Ordinarily the nature of the observations to be explained will leave no doubt as to which is correct, but in the absence of other indications the fixed response, being more parsimonious, is preferable. Means of quantitative characters can be adaptive in two different ways. The mean may represent a close approximation to a finite optimum, for example the osmotic pressure of tissue fluids. In other cases the optimum may be infinity or zero, and the observed values are merely the best that can be accomplished by selection in opposition to opposing forces and compromised by the demands of other adaptations. Measures of fleetness and of mutation rate were cited as approximations to infinity and zero. For an intermediate optimum an adequate explanation would be to show that a mutation causing deviations in either direction would be selected against. For optima at zero or infinity it is merely necessary to show that deviations in one direction would always be adversely selected. In this case there would be mechanisms designed to keep something as high or as low as possible, or at least above or below a certain threshold. There are probably many such adaptations. A primitive human instinct to choose the sweetest available fruit normally led to the eating of the ripe and nutritious, and the avoidance of the green or rancid. The same instinct now leads many of us to consume more candy than ripe fruit. A female stickleback normally is well adapted if she submits to the most active and red-bellied of her suitors. She therefore prefers grotesquely exaggerated models to those of more normal color and proportions. Such reactions to superoptimal stimuli (Tinbergen, 1951) reflect the presence of economical adaptations wherever they would function as well as more expensive ones.

Teleonomic understanding would be aided by having a recognized hierarchy of adaptations, or at least a way of specifying the subordination of one function to another. Such a system would follow Tinbergen's hierarchical classification of instincts on the basis of the nature and generality of the purposes served. The most general category would be the adaptations that are so basic as to be found in all organisms. Every organism has mechanisms for its own nutrition. Every organism has morphogenetic adaptations, those relating to growth, differentiation, reproduction, and other aspects of the completion of the life cycle. Every organism has defense mechanisms, at least in the broad sense of devices concerned with the prevention of damage to the trophic and morphogenetic machinery. Thus contractile vacuoles, eyes, and barbed spines are all defense mechanisms.

No complete explanation of a biological phenomenon can be achieved without an inquiry into its evolutionary development, and a teleonomic analysis would not proceed far without the use of historical

THE SCIENTIFIC STUDY OF ADAPTATION

data. The analysis would disclose much that is functionally inexplicable. The inversion of the retina, the crossing of the respiratory and digestive systems, and the use of the urethra for both excretory and male reproductive functions represent errors in the organization of the human body. They have no functional explanation but can be understood as aspects of functional evolution. Historical considerations are also necessary in explaining the many functionally arbitrary limitations that are always apparent in the design of an organism: Why is man a mere biped and not a Centaur? Why do marine turtles not have gills? Why must a giraffe be limited to the same number of neck vertebrae as a mouse? Lastly, evolutionary, or at least comparative data will often furnish clues to the functional meaning of biological phenomena. Pittendrigh (1958) gave some examples in his discussion of how comparisons of the reactions of two species of insects to light and moisture aided in the understanding of the function of these reactions in both species.

The specification of a general type of ecological environment, such as the Atlantic epipelagic at 60°S, specifies something about the specific problems that must be solved by the trophic, morphogenetic, and defensive adaptations of the inhabitants. Even in such a simple and homogeneous environment, however, a wide variety of approaches to the problems of life can be met. Diatoms and whales are both adapted to the same habitat, but solve their common problems with entirely different machinery, and it is in such functionally inexplicable differences between organisms that the need for evolutionary principles becomes necessary in the study of adaptation. *Whale*

THE SCIENTIFIC STUDY OF ADAPTATION

and *diatom* refer to unique historical developments. History has decreed that the diatom be an autotroph. Its approach to life is to make the most efficient possible use of inorganic salts, carbon dioxide, water, and sunlight in the manufacture of all its necessary biochemicals, and to make the most efficient possible use of these in morphogenesis. This historical decision implies the possession of a most complex and miniaturized system of enzymatic machinery, every cog of which is of vital importance. It might be of value to a diatom to have an effective sensory and motor system of defense against herbivorous animals. Any evolutionary step in this direction, however, would have meant burdening the germ plasm with additional information that would have compromised the precision of instructions concerned with enzymatic mechanisms. Such developments have been vigorously opposed by selection, because throughout the history of the diatom, even slight impairments of the enzyme systems resulted in serious reductions in fitness. The whale's ancestry was exposed to different selection pressures. Its nutrition was dependent on the efficiency of the sensory and motor mechanisms used in the capture and ingestion of animals. The synthetic enzyme systems were so compromised by the development of efficiency in predation that the whale has lost or failed to acquire the enzymes necessary to synthesize many of the necessary biochemical building blocks for its proteins, nucleic acids and coenzymes.

It is physiologically correct to say that the whale needs to ingest certain vitamins because it is unable to synthesize them, but historically, the cause-effect relationship is the other way around. The whale was able to require certain vitamins because it ordinarily ingested them. Any system will degenerate to the extent to which there is a relaxation of selection pressures for its maintenance. Emerson (1960) called attention to this principle, although his interpretation of it differs from mine, and Kosswig (1947) discussed some illustrative examples.

One of the chief goals of establishing a hierarchical organization of adaptations is to distinguish between the forces that initiated the development of an adaptation and the secondary degenerations that the adaptation, once developed, permitted. I remember a particularly relevant oral discussion of the function of inquilinism among pearlfishes. These slender-bodied fishes live in the respiratory systems of sea cucumbers. They apparently emerge at night to forage, and return at dawn to their hosts. They are largely without pigment, and there is some evidence that they are harmed by exposure to daylight. The question arose: Do these fishes enter the sea cucumbers to avoid light, or do they do so to avoid predators? The feeling of the group seemed to be that if the behavior fulfills both needs, it must be regarded as having a dual function. This is a physiologically valid conclusion, but teleonomically naïve. The two needs are surely not historically coordinate. All fish are under pressure to avoid predators, but very few are damaged by exposure to light. This must have been the condition of the ancestors of the pearlfishes. The habit of entering holothurians developed as a defense against predators, and the fish became extremely specialized in behavior and physiology for exploiting the advantages of inquilinism. This required or permitted the degeneration of a number of adaptations: the caudal fin disappeared; the eyes were reduced; and the integumentary pigments and other defenses against light were reduced in effectiveness. In this way inquilinism became a necessary part of the defense against physiological damage by light. It was not, however, as a defense against radiation that the behavior originated. The avoidance of light is a secondary need, which has arisen as a result of a degeneration of the mechanisms that ordinarily make such avoidance unnecessary.

Not all secondary needs arise by degeneration. Some are engendered as special problems to be met in the operation of a primary adaptation. If the need to avoid predators is met by taking refuge in holothurian respiratory systems, an efficient means of locating a host becomes important. A special sensory device for locating sea cucumbers would be an adaptation subordinate to inquilinism.

What would ordinarily be considered an environmental stress can become a needed resource as a result of highly effective adaptation to it. Immersion in water that is colder than the freezing point of the tissues would be a trauma for most warm-blooded animals. Large Arctic pinnipeds, however, have become so well able to maintain normal body temperatures under water at $-2°C$ that they can readily suffer heat prostration in what man would consider cold air. An environmental factor that would cause a dangerous chilling in most mammals has become a benefit to a walrus.

An even better example is provided by the adaptation of microorganisms to antibiotics. In extreme cases it may be found that a strain of bacteria requires an antibiotic that would result in the total destruction of another strain. The highly resistant bacteria apparently make obligate adjustments to the presence of the antibiotic, and when the antibiotic is not there to interact with these adjustments, the metabolism is so disrupted that normal growth cannot take place.

I believe that sleep is such a secondary requirement. In the ancestry of a species that requires sleep there may have been a population in which periodic dormancy was a facultative adaptation. It would have served to conserve energy by restricting the foraging and other vital activities to times when they could be carried out with the greatest efficiency and least danger. If nocturnal dormancy were always beneficial, however, a well-adapted facultative response system would make it a consistent and reliable feature of the life history. Adaptations that require sleep could then be added in subsequent evolutionary development, and sleep would evolve from a capacity to a requirement.

I would likewise interpret various social "needs" of animals as secondary adaptations. Allee (1940, 1951, etc.), on the basis of evidence that has subsequently been questioned (Lack, 1954A, Slobodkin, 1962), concluded that birds that breed in colonies of many pairs enjoy, as a result of the close proximity of the other individuals, a greater success in reproduction. This and other observations were taken to indicate that the need for social contacts is a basic characteristic of life and is met by gregariousness in the species noted. I would take the same observations to mean that the social species have made obligate adjustments to the presence of their fellows, and have evolved other adaptations predicated on the assumption of a certain social environment. When this environment is radically changed by the removal of the group, these adaptations may fail to function normally. I would regard it as a mistake to say that anything is a basic characteristic of life. In an organism we see only the basic characteristics of matter and the results of billions of years of adjustment to a changing environment.

I would like to nominate one more principle for initial inclusion in the science of teleonomy. This principle is that the nature of the stimuli that initiate and regulate a response may be no indication of the function of the response. This concept is certainly recognized by many biologists, but has been most clearly stated by Pittendrigh (1958). He illustrated the principle by showing that certain activities of wild populations of fruit flies are controlled by a timing mechanism regulated by visual cues from the day-night cycle of illumination. The function of the timing, however, is not an adjustment to conditions of illumination, but to changes in humidity. The sensory endowments of *Drosophila* are such that they are better able to anticipate future humidity conditions on the basis of illumination than on the basis of humidity itself. The insects adapt to an important, but poorly perceived environmental factor, humidity, by reacting to a closely correlated factor, light, that is unimportant in itself but reliably perceived. Perhaps an even better example is the timing of a plant's preparation for winter dormancy on the basis of day length. Precise observations of day length for a few days would be a better guide to what temperatures will be like two months later than would a few days of temperature observation.

THE SCIENTIFIC STUDY OF ADAPTATION

THE FORMALLY disciplined use of the theory of genic selection for problems of adaptation, as suggested in this book, should foster progress and understanding regardless of the extent to which this theory constitutes a true or adequate explanation. I am sure that by the standards of a generation hence, our current picture of evolutionary adaptation is, at best, oversimplified and naïve. It is only by the rigorous application of a theory, however, that its imperfections can be recognized and rectified. We must take the theory of natural selection in its simplest and most austere form, the differential survival of alternative alleles, and use it in an uncompromising fashion whenever a problem of adaptation arises. When such usage results in a simple and plausible explanation, the theory will thereby have demonstrated its strength. When the best such explanation is complex and not very plausible, the way is paved for a better theory.

The principle of natural selection is not, as a general rule, used by biologists in an adequately disciplined fashion. It is usually applied to problems like that of long-term morphological changes, as seen by paleontologists, or to problems of ecotypic specialization (usually climatic) and cladogenesis. These phenomena make easy demands on a theory of adaptation. Most of the conclusions on patterns of speciation would be much the same whether based on Lamarckian, nineteenth-century Darwinian, or modern genetic concepts. The fact that a modern paper on speciation in a certain genus contains such terms as mutation, gene flow, and selection, need not imply that it is conceptually much advanced beyond what Lamarck or Darwin might have written. Darwin's and even Lamarck's concepts form a perfectly ade-

THE SCIENTIFIC STUDY OF ADAPTATION

quate basis for explaining most of the phenomena of systematics.

I would certainly not suggest that any field of scientific investigation should be pursued with decreased effort, and I would urge that the field of evolutionary systematics is of high importance. Nevertheless, I maintain that such studies are not likely to lead to important advances in achieving a general understanding of evolutionary adaptation. The same conclusion was cogently advanced by Epling and Catlin (1950).

The important advances will come from quantitative studies of the phenomena of adaptation, not studies merely of the superficial ecotypic adaptation emphasized in systematics, but studies of the distribution and phylogenetic variation in the general strategies employed in the game of life. Darwin devoted a considerable amount of effort and space in his publications to such problems. Despite his statement that the origin of species is the "problem of problems" in natural history, he dealt with much besides cladogenesis and descriptive phylogeny in *The Origin of Species*. In this book and elsewhere he devoted much space to explanations for the origin and evolution of such adaptations as sexuality, intelligence, aerial flight, organs of extreme perfection, such as the vertebrate eye, those of seemingly trivial importance, such as fly-swatter tails, and the group-related adaptations of the social insects. Darwin's works show a more wholesome balance of topics than is shown by the modern evolutionary literature, with its bias for climatic adaptation and cladogenesis.

I believe that important insights can still be obtained from theoretical studies of some of the same

THE SCIENTIFIC STUDY OF ADAPTATION

questions that Darwin discussed in 1859. Similar attention should be devoted to population sex ratios; the significance of the X–Y sex determining mechanism and the seemingly arbitrary distribution of male and female heterogamety; the significance of chromosome number and linkage relations; phylogenetic variation in reproductive physiology and behavior; phylogenetic variation in life cycles in general, and the adaptive significance of paedogenesis, parthenogenesis, apomixis, metagenesis, metamorphosis, etc.; phylogenetic and ontogenetic distribution of developmental rates, especially the long juvenile stages of birds and the presence and duration of pelagic dispersal stages of sedentary marine organisms; the origin of any really outstanding characters such as human intelligence and insect societies; and the evolutionary loss of adaptations of all kinds in groups that once possessed them.

I would regard the problem of sex ratio as solved (see pp. 146-156). There is, of course, a large absolute mass of relevant literature on all the other problems mentioned. The attention devoted to them, however, is minute in comparison with the effort expended on taxonomic problems that are of no greater importance.

There may be an apt analogy between the theory of natural selection as it is today and the atomic theory of two centuries ago. The concept of matter as fundamentally particulate had been used in an undisciplined fashion at least since the days of Democritus. It was invoked, as natural selection often is today, whenever it seemed expedient to do so, but nothing was ever really demanded of the theory. Real tests, such as the prediction of temperature-volume relationships in gases or of the weights of the products of a chemical reaction would have been impossible to formulate. The theory had to be stated in an explicit, quantitative, and uncompromising form before it was possible to recognize logical implications or to demand that there be a precise congruence between the theory and observation. The essential service was provided by Dalton, who proposed six theoretical postulates about the nature of atoms. To Dalton, atoms were always thus and so. His statements allowed no compromises and took no refuge in vagueness. Inconsistencies with observational data very soon came to light, and after two peaceful but largely fruitless millennia the concept of material atomism was seriously brought into question. The theory survived in a modified form, even though every one of Dalton's six postulates turned out to be wrong, or at least inaccurate. Dalton performed an enormous service, because his theory provided a basis for questions that could be decided by objective evidence. In this way he helped to open the door to the modern era of the science of chemistry.

Perhaps today's theory of natural selection, which is essentially that provided more than thirty years ago by Fisher, Haldane, and Wright, is somewhat like Dalton's atomic theory. It may not, in any absolute or permanent sense, represent the truth, but I am convinced that it is the light and the way.

Literature Cited

ALLEE, W. C., 1931, *Animal Aggregations: A Study in General Sociology*, University of Chicago Press, ix, 431 pp.

——, 1940, Concerning the origin of sociality in animals, *Scientia* 1940:154-160.

——, 1943, Where angels fear to tread: a contribution from general sociology to human ethics, *Science* 97:517-525.

——, 1951, *Cooperation among Animals*, New York, Henry Schuman, 233 pp.

ALLEE, W. C., ALFRED E. EMERSON, ORLANDO PARK, THOMAS PARK, KARL P. SCHMIDT, 1949, *Principles of Animal Ecology*, Philadelphia, W. B. Saunders Co., xii, 837 pp.

ALLISON, A. C., 1955, Aspects of polymorphism in man, *Cold Spring Harbor Symp. Quant. Biol.* 20:239-255.

ALTMAN, STUART A., 1962, A field study of the sociobiology of rhesus monkeys, *Macaca mulatta*, *Ann. N. Y. Acad. Sci.* 102:338-435.

AMADON, DEAN, 1959, The significance of sexual differences in size among birds, *Proc. Am. Phil. Soc.* 103:531-536.

——, 1964, The evolution of low reproductive rates in birds, *Evolution* 18:105-110.

ANDERSEN, F. SØGAARD, 1961, Effect of density on animal sex ratio, *Oikos* 12:1-16.

ANDERSON, EDGAR, 1953, Introgressive hybridization, *Biol. Rev. Cambridge Phil. Soc.* 28:280-307.

AUERBACH, C., 1956, *Genetics in the Atomic Age*, Edinburgh, Oliver & Boyd, 106 pp.

BARKER, J. S. F., 1963, The estimation of relative fitness of *Drosophila* populations, II, Experimental evaluation of factors affecting fitness, *Evolution* 17:56-71.

BARNES, H., 1962, So-called anecdysis in *Balanus balanoides* and the effect of breeding upon the growth of calcareous shell of some common barnacles, *Limnol. Oceanog.* 7:462-473.

LITERATURE CITED

BARNEY, R. L., B. J. ANSON, 1920, Life history and ecology of the pygmy sunfish, *Elassoma zonatum*, Ecology 1:241-256.

BATEMAN, A. J., 1949, Analysis of data on sexual selection, *Evolution* 3:174-177.

BERGERARD, J., 1962, Parthenogenesis in the Phasmidae, *Endeavor* 21:137-143.

BIRCH, L. C., 1957, The meanings of competition, *Am. Naturalist* 91:5-18.

———, 1960, The genetic factor in population ecology, *Am. Naturalist* 94:5-24.

BLOOD, DONALD A., 1963, Some aspects of behavior in a bighorn herd, *Can. Field Naturalist* 77:77-94.

BLUM, HAROLD F., 1961, Does the melanin pigment of human skin have adaptive value? An essay in human ecology and the evolution of the race, *Quart. Rev. Biol.* 36:50-63.

———, 1963, On the origin and evolution of human culture, *Am. Scientist* 51:32-37.

BODMER, W. F., A. W. F. EDWARDS, 1960, Natural selection and the sex ratio, *Ann. Human Genet.* 24:239-244.

BONNER, JOHN TYLER, 1957, A theory of the control of differentiation in the cellular slime molds, *Quart. Rev. Biol.* 32: 232-246.

———, 1958, *The Evolution of Development*, Cambridge University Press, 102 pp.

BORMANN, F. H., 1962, Root grafting and non-competitive relationships between trees, pp. 237-246 in: *Tree Growth*, T. T. Kozlowski, ed., New York, Ronald Press, xi, 442 pp.

BORRADAILE, L. A., F. A. POTTS, L. E. S. EASTHAM, J. T. SAUNDERS, G. A. KERKUT, 1961, *The Invertebrata*, Cambridge University Press, xviii, 820 pp.

BOYDEN, ALAN A., 1953, Comparative evolution with special reference to primitive mechanisms, *Evolution* 7:21-30.

———, 1954, The significance of asexual reproduction, *Syst. Zool.* 3:26-37, 47.

BRAESTRUP, F. W., 1963, The function of communal displays, *Dansk Ornithol. Foren. Tidsskr.* 57:133-142.

BREDER, CHARLES M., 1936, The reproductive habits of North American sunfishes (family Centrarchidae), *Zoologica* 21: 1-48.

LITERATURE CITED

——, 1952, On the utility of the saw of the sawfish, *Copeia* 1952:90-91.

——, 1959, Studies on social groupings in fishes, *Bull. Am. Mus. Nat. Hist.* 117:395-481, pls. 70-80.

BRERETON, J. LE GAY, 1962A, Evolved regulatory mechanisms of population control, pp. 81-93 in: *The Evolution of Living Organisms*, G. W. Leeper, ed., Melbourne University Press, xi, 459 pp.

——, 1962B, A laboratory study of population regulation in *Tribolium confusum*, *Ecology* 43:63-69.

BROCK, VERNON E., ROBERT H. RIFFENBURGH, 1960, Fish schooling: a possible factor in reducing predation, *J. Conseil, Conseil Perm. Intern. Exploration Mer* 25:307-317.

BROWN, WILLIAM L., JR., 1958, General adaptation and evolution, *Syst. Zool.* 7:157-168.

BUDD, G. M., 1962, Population studies in rookeries of the emperor penguin *Aptenodytes forsteri*, *Proc. Zool. Soc. London* 139:365-388, 1 pl.

BULLIS, HARVEY R., JR., 1960, Observations on the feeding behavior of white-tip sharks on schooling fishes, *Ecology* 42:194-195.

BURKHOLDER, PAUL R., 1952, Cooperation and conflict among primitive organisms, *Am. Scientist* 40:601-631.

BURNET, F. M., 1961, Immunological recognition of self, *Science* 133: 307-311.

——, 1962, *The Integrity of the Body*, Harvard University Press, 189 pp.

BURTT, E. A., 1955, *The Teachings of the Compassionate Buddha*, New York, Mentor, MD 131, 247 pp.

BUZZATI-TRAVERSO, A. A., 1954, On the role of mutation rate in evolution, *Caryologia* 6(suppl.):450-462.

CAGLE, FRED R., 1955, Courtship behavior in juvenile turtles, *Copeia* 1955:307.

CARLISLE, D. B., 1962, On the venom of the lesser weeverfish *Trachinus vipera*, *J. Marine Biol. Assoc. U. K.* 42:155-162.

CARSON, HAMPTON L., 1961, Heterosis and fitness in experimental populations of *Drosophila melanogaster*, *Evolution* 15:496-509.

LITERATURE CITED

CATCHESIDE, D. G., 1951, *The Genetics of Micro-organisms*, London, Pitman, vii, 223 pp.

CLARKE, C. A., C. G. C. DICKSON, P. M. SHEPPARD, 1963, Larval color pattern in *Papilio demodocus*, *Evolution* 17: 130-137.

CLARKE, GEORGE L., 1954, *Elements of Ecology*, New York, Wiley, xiv, 534 pp.

COLE, LAMONT C., 1954, The population consequences of life history phenomena, *Quart. Rev. Biol.* 29:103-137.

———, 1958, Sketches of general and comparative demography, *Cold Spring Harbor Symp. Quant. Biol.* 22:1-15.

COMFORT, ALEX, 1956, *The Biology of Senescence*, New York, Rinehart & Co., xiii, 257 pp.

CORRENS, C., 1927, Der Unterschied in der Keimungsgeschwindigkeit der männchensamen und weibchensamen bei *Melandrium*, *Hereditas* 9:33-44.

COTT, HUGH B., 1954, Allaesthetic selection and its evolutionary aspects, pp. 47-70 in: *Evolution as a Process*, J. S. Huxley, A. C. Hardy, E. B. Ford, ed., London, Allen & Unwin, 367 pp.

CRISP, D. J., BHUPENDRA PATEL, 1961, The interaction between breeding and growth rate in the barnacle *Elminius modestus* Darwin, *Limnol. Oceanog.* 6:105-115.

CROSBY, J. L., 1963, The evolution and nature of dominance, *J. Theoret. Biol.* 5:35-51.

CULLEN, E., 1957, Adaptations in the kittiwake to cliffnesting, *Ibis* 99:275-302.

DARLING, F. FRASER, 1938, *Bird Flocks and the Breeding Cycle*, Cambridge University Press, x, 124 pp.

DARLINGTON, C. D., 1958, *The Evolution of Genetic Systems*, New York, Basic Books, xi, 265 pp.

DARLINGTON, C. D., KENNETH MATHER, 1949, *The Elements of Genetics*, London, Allen & Unwin, 446 pp.

DARWIN, CHARLES R., 1882, *The Variation of Animals and Plants under Domestication*, London, John Murray, vol. I, xiv, 472 pp., vol. II, x, 495 pp.

———, 1896, *The Descent of Man and Selection in Relation to Sex*, New York, D. Appleton & Co., xvi, 688 pp.

LITERATURE CITED

DIJKGRAAF, V. S., 1952, Bau und Funktionen der Seitenorgane und des Ohrlabyrinthes bei Fischen, *Experientia* 8:205-216.

——, 1963, The functioning and significance of the lateral-line organs, *Biol. Rev. Cambridge Phil. Soc.* 38:51-105.

DOBZHANSKY, THEODOSIUS, 1951, *Genetics and the Origin of Species*, Columbia University Press, xiv, 364 pp.

——, 1959, Evolution of genes and genes in evolution, *Cold Spring Harbor Symp. Quant. Biol.* 24:15-30.

——, 1963, Genetics of natural populations, XXXIII, A progress report on genetic changes in populations of *Drosophila pseudoobscura* and *Drosophila persimilis* in a locality in California, *Evolution* 17:333-339.

DOBZHANSKY, THEODOSIUS, M. F. A. MONTAGU, 1947, Natural selection and the mental capacities of mankind, *Science* 106:587-590.

DOUGHERTY, ELLSWORTH C., 1955, Comparative evolution and the origin of sexuality, *Syst. Zool.* 4:145-169.

DUNBAR, M. J., 1960, The evolution of stability in marine environments; natural selection at the level of the ecosystem, *Am. Naturalist* 94:129-136.

EDWARDS, A. W. F., 1960, Natural selection and the sex ratio, *Nature* 188:960-961.

EHRENSVÄRD, GÖSTA, 1962, *Life: Its Origin and Development*, Minneapolis, Burgess, 204 pp.

EHRMAN, LEE, 1963, Hybrid sterility as an isolating mechanism in the genus *Drosophila*, *Quart. Rev. Biol.* 37:279-302.

ELTON, C., 1942, *Voles, Mice and Lemmings. Problems in Population Dynamics*, Oxford, Clarendon Press, 496 pp.

EMERSON, ALFRED E., 1960, The evolution of adaptation in population systems, pp. 307-348 in: *Evolution after Darwin*, vol. 1, Sol Tax, ed., University of Chicago Press, viii, 629 pp.

——, 1961, Vestigial characters of termites and processes of regressive evolution, *Evolution* 15:115-131.

EPLING, CARL, WESLEY CATLIN, 1950, The relation of taxonomic method to an explanation of evolution, *Heredity* 4:313-325.

ESSIG, E. O., 1942, *College Entomology*, New York, Macmillan, vii, 900 pp.

LITERATURE CITED

EVANS, L. T., ed., 1962, *Environmental Control of Plant Growth*, New York, Academic Press, 467 pp.

FELIN, FRANCES E., 1951, Growth characteristics of the Poeciliid fish, *Platypoecilus maculatus*, *Copeia* 1951:15-28.

FIEDLER, KURT, 1954, Vergleichende Verhaltensstudien an Seenadeln, Schlangennadeln und Seepferdchen (Syngnathidae), *Z. Tierpsychol.* 11: 358-416.

FILOSA, M. F., 1962, Heterocytosis in cellular slime molds, *Am. Naturalist* 96:79-92.

FINK, BERNARD D., 1959, Observation of porpoise predation on a school of Pacific sardines, *Calif. Fish. Game* 45:216-217.

FISHER, JAMES, 1954, Evolution and bird sociality, pp. 71-83 in: *Evolution as a Process*, J. S. Huxley, A. C. Hardy, E. B. Ford, eds., London, Allen & Unwin, 367 pp.

FISHER, RONALD A., 1930, *The Genetical Theory of Natural Selection*, Oxford, Clarendon Press; reprinted 1958, New York, Dover, xiv, 291 pp.

——, 1954, Retrospect of the criticisms of the theory of natural selection, pp. 84-98 in: *Evolution as a Process*, J. S. Huxley, A. C. Hardy, E. B. Ford, eds., London, Allen & Unwin, 367 pp.

FISHER, RONALD A., E. B. FORD, 1947, The spread of a gene in natural conditions in a colony of the moth, *Panaxia dominula* (L), *Heredity* 1:143-174.

FORD, E. B., 1956, Rapid evolution and the conditions which make it possible, *Cold Spring Harbor Symp. Quant. Biol.* 20:230-238.

FOWLER, JAMES A., 1961, Anatomy and development of racial hybrids of *Rana pipiens*, *J. Morphol.* 109:251-268.

FRAENKEL, GOTTFRIED S., 1959, The *raison d'être* of secondary plant substances, *Science* 129:1466-1470.

FRANK, FRITZ, 1957, The causality of microtine cycles in Germany, *J. Wildlife Management* 21:113-121.

FREEDMAN, LAWRENCE Z., ANNE ROE, 1958, Evolution and human behavior, pp. 455-479 in: *Behavior and Evolution*, A. Roe, G. G. Simpson, eds., Yale University Press, viii, 557 pp.

LITERATURE CITED

GOTTO, R. V., 1962, Egg number and ecology in commensal and parasitic copepods, *Ann. Mag. Nat. Hist.* 13S, 5:97-107.

GUHL, A. M., W. C. ALLEE, 1944, Some measurable effects of social organization in flocks of hens, *Physiol. Zool.* 17: 320-347.

HAARTMAN, LARS VON, 1957, Adaptation in hole-nesting birds, *Evolution* 11:339-347.

HAGAN, H. R., 1951, *Embryology of the Viviparous Insects*, New York, Ronald Press, xiv, 472 pp.

HALDANE, J. B. S., 1931, A mathematical theory of natural and artificial selection, Part VII, Selection intensity as a function of mortality rate, *Proc. Cambridge Phil. Soc.* 27:131-142.

———, 1932, *The Causes of Evolution*, London, Longmans, vii, 235 pp.

HALL, E. RAYMOND, KEITH R. KELSON, 1959, *The Mammals of North America*, New York, Ronald Press, vol. 1, pp. xxx, 1-546, 1-79; vol. 2, pp. viii, 547-1083, 1-79.

HALL, K. R. L., 1960, Social vigilance behavior of the chacma baboon, *Papio ursinus*, *Behavior* 16:261-294.

HALSTEAD, BRUCE W., 1959, *Dangerous Marine Animals*, Cambridge, Md., Cornell Maritime Press, ix, 146 pp.

HALSTEAD, BRUCE W., F. RENE MODGLIN, 1950, A preliminary report on the venom apparatus of the bat-ray, *Holorhinus californicus*, *Copeia* 1950:165-175.

HAMILTON, W. D., 1964A, The genetical evolution of social behaviour, I, *J. Theoret. Biol.* 7:1-16.

———, 1964B, The genetical evolution of social behaviour, II, *J. Theoret. Biol.* 7:17-52.

HARPER, JOHN, L., 1960, Factors controlling plant numbers, pp. 119-132 in: *The Biology of Weeds*, John L. Harper, ed., Oxford, Blackwell, xv, 256 pp.

HARRINGTON, R. W., 1948, The life cycle and fertility of the bridled shiner, *Notropis bifrenatus* (Cope), *Am. Midland Naturalist* 39:83-92.

HIRAIZUMI, YUICHIRO, L. SANDLER, JAMES F. CROW, 1960, Meiotic drive in natural populations of *Drosophila melano-*

LITERATURE CITED

gaster, III, Populational implications of the segregation-distorter locus, *Evolution* 14:433-444.

HOCHMAN, BENJAMIN, 1961, Isoallelic competition in populations of *Drosophila melanogaster* containing a genetically heterogeneous background, *Evolution* 15:239-246.

HODDER, V. M., 1963, Fecundity of Grand Bank haddock, *J. Fisheries Res. Board Can.* 20:1465-1487.

HUBBS, CARL L., 1955, Hybridization between fish species in nature, *Syst. Zool.* 4:1-20.

HUBBS, CLARK, 1958, Geographic variations in egg complement of *Percina caprodes* and *Etheostoma spectabile*, *Copeia* 1958:102-105.

HUXLEY, JULIAN S., 1942, *Evolution, the Modern Synthesis*, New York, Harper, 645 pp.

———, 1953, *Evolution in Action*, New York, Harper, x, 182 pp.

———, 1954, The evolutionary process, pp. 1-23 in: *Evolution as a Process*, J. Huxley, A. C. Hardy, E. B. Ford. eds., London, Allen & Unwin, 367 pp.

———, 1958, Cultural process and evolution, Chap. 20 in: *Behavior and Evolution*, A. Roe, G. G. Simpson, eds., Yale University Press, viii, 557 pp.

IVES, P. T., 1950, The importance of mutation rate genes in evolution, *Evolution* 4:236-252.

JONES, J. W., H. B. N. HYNES, 1950, The age and growth of *Gasterosteus aculeatus, Pygosteus pungitius* and *Spinachia vulgaris*, as shown by their otoliths, *J. Animal Ecol.* 19:59.

KENDEIGH, S. CHARLES, 1952, Parental care and its evolution in birds, *Illinois Biol. Monog.* 22:1-356.

KIMURA, MOTOO, 1956, A model of a genetic system which leads to closer linkage by natural selection, *Evolution* 10: 278-287.

———, 1958, On the change of population fitness by natural selection, *Heredity* 12:145-167.

———, 1960, Optimum mutation rate and degree of dominance as determined by the principle of minimum genetic load, *J. Genet.* 57:21-34.

LITERATURE CITED

———, 1961, Natural selection as the process of accumulating genetic information in adaptive evolution, *Genet. Res.* 2: 127-140.

KLAUBER, LAURENCE M., 1956, *Rattlesnakes: Their Habits, Life Histories, and Influence on Mankind*, University of California Press, vol. 1, pp. xxix, 1-708; vol. 2, pp. xvii, 709-1476.

KNIGHT-JONES, E. W., J. MOYSE, 1961, Intraspecific competition in sedentary marine animals, *Symp. Soc. Exp. Biol.* 15:72-95.

KOFORD, CARL B., 1957, The vicuna and the Puna, *Ecol. Monog.* 27:153-219.

KOSSWIG, CURT, 1946, Bemerkungen zur degenerativen Evolution, *Compt. Rend. Ann. Arch. Soc. Turq. Sci. Phys. Nat.* 12:135-162.

LACK, DAVID, 1954A, *The Natural Regulation of Animal Numbers*, Oxford University Press, viii, 343 pp.

———, 1954B, The evolution of reproductive rates, pp. 143-156 in: *Evolution as a Process*, J. S. Huxley, A. C. Hardy, E. B. Ford, eds., London, Allen & Unwin, 367 pp.

LAGLER, KARL F., JOHN E. BARDACH, ROBERT R. MILLER, 1962, *Ichthyology*, New York, Wiley, xiii, 545 pp.

LEOPOLD, A. C., 1961, Senescence in plant development, *Science* 134:1727-1732.

LERNER, I. MICHAEL, 1953, *Genetic Homeostasis*, New York, Wiley, vii, 134 pp.

LEVENE, HOWARD, OLGA PAVLOVSKY, THEODOSIUS DOBZHANSKY, 1958, Dependence of the adaptive values of certain genotypes in *Drosophila pseudoobscura* on the composition of the gene pool, *Evolution* 12:18-23.

LEVITAN, MAX, 1961, Proof of an adaptive linkage association, *Science* 134:1617-1619.

LEWIS, D., 1942, The evolution of sex in flowering plants, *Biol. Rev. Cambridge Phil. Soc.* 17:46-67.

LEWIS, D., LESLIE K. CROWE, 1956, The genetics and evolution of gynodioecy, *Evolution* 10:115-125.

LEWONTIN, R. C., 1958A, Studies on heterozygosity and homeostasis, II, Loss of heterosis in a constant environment, *Evolution* 12:494-503.

LITERATURE CITED

——, 1958B, The adaptations of populations to varying environments, *Cold Spring Harbor Symp. Quant. Biol.* 22: 395-408.

——, 1961, Evolution and the theory of games, *J. Theoret. Biol.* 1:382-403.

——, 1962, Interdeme selection controlling a polymorphism in the house mouse, *Am. Naturalist* 96:65-78.

LEWONTIN, R. C., L. C. DUNN, 1960, The evolutionary dynamics of a polymorphism in the house mouse, *Genetics* 45:705-722.

LEWONTIN, R. C., KEN-ICHI KOJIMA, 1960, The evolutionary dynamics of complex polymorphisms, *Evolution* 14:458-472.

LI, C. C., 1955, *Population Genetics*, University of Chicago Press, xi, 366 pp.

LIDICKER, WILLIAM Z., 1962, Emigration as a possible mechanism permitting the regulation of population density below carrying capacity, *Am. Naturalist* 96:29-33.

LYDECKKER, R., 1898, *Wild Oxen, Sheep, and Goats of All Lands*, London, Rowland Ward, xiv, 318 pp.

MCCLINTOCK, BARBARA, 1951, Chromosome organization and genic expression, *Cold Spring Harbor Symp. Quant. Biol.* 16:13-46.

MAKINO, SAJIRO, 1951, *An Atlas of the Chromosome Numbers in Animals*, Iowa State University Press, xxviii, 290 pp.

MATHER, KENNETH, 1953, The genetical structure of populations, *Symp. Soc. Exp. Biol.* 7:66-95.

——, 1961, Competition and cooperation, *Symp. Soc. Exp. Biol.* 15:264-281.

MAYR, ERNST, 1954, Change of genetic environment and evolution, pp. 157-180 in: *Evolution as a Process*, J. S. Huxley, A. C. Hardy, E. B. Ford, eds., London, Allen & Unwin, 367 pp.

——, 1963, *Animal Species and Their Evolution*, Harvard University Press, 813 pp.

MEDAWAR, P. B., 1952, *An Unsolved Problem in Biology*, London, H. K. Lewis, 24 pp.

LITERATURE CITED

———, 1960, *The Future of Man*, New York, Basic Books, 128 pp.

———, 1961, Immunological tolerance, *Science* 133:303-306.

MICHIE, DONALD, 1958, The third stage in genetics, pp. 56-84 in: *A Century of Darwin*, S. A. Barnett, ed., London, Heinemann, xvi, 376 pp.

MILNE, A., 1961, Definition of competition among animals, *Symp. Soc. Exp. Biol.* 15:40-61.

MIRSKY, A. E., HANS RIS, 1951, The desoxyribonucleic acid content of animal cells and its evolutionary significance, *J. Gen. Physiol.* 34:451-462.

MONTAGU, M. F. ASHLEY, 1952, *Darwin, Competition and Cooperation*, New York, Henry Schuman, 148 pp.

MORRIS, DESMOND, 1955, The causation of pseudofemale and pseudomale behavior: a further comment, *Behavior* 8:46-56.

MOTTRAM, J. C., 1915, The distribution of secondary sexual characters amongst birds, with relation to their liability to the attack of enemies, *Proc. Zool. Soc. London* 7:663-678.

MULLER, H. J., 1948, Evidence of the precision of genetic adaptation, *Harvey Lectures* 43:165-229.

MURIE, OLAUS J., 1935, Alaska-Yukon Caribou, *U. S. Bur. Biol. Surv. North American Fauna* 55:1-93.

MURPHY, R. C., 1936, *Oceanic Birds of South America*, American Museum of Natural History, 2 vols., xx, 1245 pp.

MYERS, GEORGE S., 1952, Annual fishes, *Aquarium J.* 23:125-141.

NEEDHAM, A. E., 1952, *Regeneration and Wound Healing*, New York, Wiley, viii, 152 pp.

NICHOLSON, J. A., 1956, Density governed reaction, the counterpart of selection in evolution, *Cold Spring Harbor Symp. Quant. Biol.* 20:288-293.

———, 1960, The role of population dynamics in natural selection, pp. 477-521 in: *Evolution after Darwin*, vol. 1, Sol Tax, ed., University of Chicago Press, viii, 629 pp.

NIKOLSKY, G. V., 1962, *The Ecology of Fishes*, New York, Academic Press, xv, 352 pp.

NOBLE, G. KINGSLEY, 1931, *The Biology of the Amphibia*, New York, Dover Reprint (1954), 577 pp.

LITERATURE CITED

NORMAN, J. R., 1949, *A History of Fishes*, New York, A. A. Wyn, xv, 463 pp.

O'DONALD, P., 1962, The theory of sexual selection, *Heredity* 17:541-552.

ODUM, H. T., W. C. ALLEE, 1956, A note on the stable point of populations showing both interspecific cooperation and disoperation, *Ecology* 35:95-97.

OGLE, KENNETH N., 1962, The visual space sense, *Science* 135:763-771.

PALEY, WILLIAM, 1836, *Natural Theology*, vol. 1, London, Charles Knight, xv, 456 pp.

PARK, THOMAS, MONTE LLOYD, 1955, Natural selection and the outcome of competition, *Am. Naturalist* 89:235-240.

PENNY, RICHARD L., 1962, Voices of the adélie, *Nat. Hist.* 71:16-26.

PIMENTEL, DAVID, 1961, Animal population regulation by the genetic feedback mechanism, *Am. Naturalist* 95:65-79.

PITTENDRIGH, COLIN S., 1958, Adaptation, natural selection, and behavior, Chap. 18 (pp. 390-416) in: *Behavior and Evolution*, A. Roe, G. G. Simpson, eds., Yale University Press, viii, 557 pp.

RAND, AUSTIN L., 1954, Social feeding behavior of birds, *Fieldiana Zool.* 36:1-71.

RATTENBURY, J. A., 1962, Cyclic hybridization as a survival mechanism in the New Zealand forest flora, *Evolution* 16:348-363.

REED, T. E., 1959, The definition of relative fitness of individuals with specific genetic traits, *Am. J. Human Genet.* 11:137-155.

RICH, WALTER H., 1947, The swordfish and swordfishery of New England, *Proc. Portland Soc. Nat. Hist.* 4:5-102.

RICHDALE, L. E., 1951, *Sexual Behavior in Penguins*, University of Kansas Press, xi, 316 pp.

———, 1957, *A Population Study of Penguins*, Oxford, Clarendon Press, 195 pp., 2 pls.

RITTER, WILLIAM E., 1938, *The California Woodpecker and I*, University of California Press, xiii, 340 pp.

LITERATURE CITED

Ross, Herbert H., 1962, *A Synthesis of Evolutionary Theory*, Englewood Cliffs, Prentice Hall, ix, 387 pp.

Russell, E. S., 1945, *The Directiveness of Organic Activities*, Cambridge University Press, viii, 196 pp.

Salt, George, 1961, Competition among insect parasitoids, *Symp. Soc. Exp. Biol.* 15:96-119.

Sandler, L., E. Novitski, 1957, Meiotic drive as an evolutionary force, *Am. Naturalist* 91:105-110.

Schmidt, Karl P., Robert F. Inger, 1957, *Living Reptiles of the World*, New York, Doubleday, 287 pp.

Shaw, Richard F., 1958, The theoretical genetics of the sex ratio, *Genetics* 47:149-163.

Sheppard, P. M., 1954, Evolution in bisexually reproducing organisms, pp. 201-218 in: *Evolution as a Process*, J. S. Huxley, A. C. Hardy, E. B. Ford, eds., London, Allen & Unwin, 367 pp.

———, 1958, *Natural Selection and Heredity*, London, Hutchinson, 212 pp.

Simpson, George Gaylord, 1944, *Tempo and Mode in Evolution*, Columbia University Press, xiii, 237 pp.

———, 1953, *The Major Features of Evolution*, Columbia University Press, xx, 434 pp.

———, 1962, Biology and the nature of life, *Science* 139:81-88.

Singer, Ronald, 1962, Emerging man in Africa, *Nat. Hist.* 71:11-21.

Skutch, Alexander F., 1961, Helpers among birds, *Condor* 63:198-226.

Slijper, E. J., A. J. Pomerans (transl.), 1962, *Whales*, New York, Basic Books, 475 pp.

Slobodkin, L. Basil, 1953, An algebra of population growth, *Ecology* 34:513-519.

———, 1954, Population dynamics of *Daphnia obtusa* Kurz, *Ecol. Monog.* 24:69-88.

———, 1959, A laboratory study of the effect of removal of newborn animals from a population, *Proc. Natl. Acad. Sci. U.S.* 43:780-782.

———, 1962, *Growth and Regulation of Animal Populations*, New York, Holt, Reinhart, & Winston, vii, 184 pp.

Slobodkin, L. Basil, S. Richman, 1956, The effect of re-

LITERATURE CITED

moval of fixed percentages of newborn on size and variability in populations of *Daphnia pulicaria* (Forbes), *Limnol. Oceanog.* 1:209-237.

SMITH, J. L. B., 1951, A case of poisoning by the stonefish. *Synanceja verrucosa, Copeia* 1951:207-210.

SMITH, J. MAYNARD, 1958, Sexual selection, pp. 231-244 in: *A Century of Darwin*, S. A. Barnett, ed., London, Heinemann, xvi, 376 pp.

SNYDER, ROBERT L., 1961, Evolution and integration of mechanisms that regulate population growth, *Proc. Natl. Acad. Sci. U.S.* 47:449-455.

SOMMERHOFF, G., 1950, *Analytical Biology*, Oxford University Press, viii, 207 pp.

SPIETH, HERMAN T., 1958, Behavior and isolating mechanisms, Chap. 17 (pp. 363-389) in: *Behavior and Evolution*, A. Roe, G. G. Simpson, eds., Yale University Press, viii, 557 pp.

STALKER, HARRISON D., 1956, On the evolution of parthenogenesis in Lonchoptera (Diptera), *Evolution* 10:345-359.

STEBBINS, G. LEDYARD, 1960, The comparative evolution of genetic systems, pp. 197-226 in: *Evolution after Darwin*, vol. 1, Sol Tax, ed., University of Chicago Press, viii, 629 pp.

SUOMALAINEN, E., 1953, Parthenogenesis in animals, *Advan. Genetics* 3:193-253.

SVÄRDSON, GUNNAR, 1949, Natural selection and egg number in fish, *Rept. Inst. Freshwater Res., Drottningholm* 29:115-122.

THODAY, J. M., 1953, Components of fitness, *Symp. Soc. Exp. Biol.* 1:96-113.

——, 1958, Natural selection and biological progress, pp. 313-333 in: *A Century of Darwin*, S. A. Barnett, ed., London, Heinemann, xvi, 376 pp.

THOMPSON, D. Q., 1955, The 1953 lemming emigration at Point Barrow, Alaska, *Arctic* 8:37-45.

TINBERGEN, N., 1951, *The Study of Instinct*, Oxford University Press, 228 pp.

——, 1957, The functions of territory, *Bird Study* 4:14-27.

LITERATURE CITED

UNDERWOOD, GARTH, 1954, Categories of adaptation, *Evolution* 8:365-377.

VENDRELY, R., 1955, The desoxyribonucleic acid content of the nucleus, pp. 155-180 in: *The Nucleic Acids*, vol. 2, E. Chargaff, J. N. Davidson, eds., New York, Academic Press, xi, 576 pp.

VORONTSOVA, M. A., L. D. LIOSNER, 1960, *Asexual Propagation and Regeneration*, New York, Pergamon Press, 489 pp.

WADDINGTON, C. H., 1956, Genetic assimilation of the *Bithorax* phenotype, *Evolution* 10:1-13.

———, 1957, *The Strategy of the Genes*, London, Allen & Unwin, ix, 262 pp.

———, 1958, Theories of evolution, pp. 1-18 in: *A Century of Darwin*, S. A. Barnett, ed., London, Heinemann, xvi, 376 pp.

———, 1959, Evolutionary adaptation, *Perspectives Biol. Med.* 2:379-401.

———, 1961, *The Nature of Life*, London, Allen & Unwin, 131 pp.

———, 1962, *New Patterns in Genetics and Development*, Columbia University Press, xiv, 271 pp.

WARBURTON, FREDERICK E., 1955, Feedback in development and its evolutionary significance, *Am. Naturalist* 89:129-140.

WARDLAW, C. W., 1955, *Embryogenesis in Plants*, London, Methuen, ix, 381 pp.

WEISMANN, A., 1882, The duration of life, Chap. 1 (vol. 1) in: *Essays upon Heredity and Kindred Biological Problems*, Oxford University Press, xv, 471 pp.

———, 1904, *The Evolution Theory*, London, Arnold, vol. 1, xvi, 416 pp.; vol. 2, iii, 405 pp.

WELLENSIEK, UTE, 1953, Die Allometrieverhältnisse und Konstruktionsänderung bei dem kleinsten Fisch im Vergleich mit etwas grösseren verwandten Formen, *Jahrb. Abt. Anat. Ontog. Tiere* 73:187-228.

WHITE, M. J. D., LESLEY E. ANDREW, 1962, Effects of chromosomal inversions on size and relative viability in the grasshopper *Moraba scurra*, pp. 94-101 in: *The Evolution*

LITERATURE CITED

of *Living Organisms*, G. W. Leeper, ed., Melbourne University Press, xi, 459 pp.

WILLIAMS, GEORGE C., 1957, Pleiotropy, natural selection, and the evolution of senescence, *Evolution* 11:398-411.

———, 1959, Ovary weights of darters: a test of the alleged association of parental care with reduced fecundity in fishes, *Copeia* 1959:18-24.

———, 1964, Measurement of consociation among fishes and comments on the evolution of schooling, *Michigan State Univ. Mus. Publ., Biol. Ser.* 2:351-383.

WILLIAMS, GEORGE C., DORIS C. WILLIAMS, 1957, Natural selection of individually harmful social adaptations among sibs with special reference to social insects, *Evolution* 11: 32-39.

WILSON, EDWARD O., 1963, Social modifications related to rareness in ant species, *Evolution* 17:249-253.

WRIGHT, SEWALL, 1931, Evolution in Mendelian populations, *Genetics* 16:97-159.

———, 1945, Tempo and mode in evolution: a critical review, *Ecology* 26:415-419.

———, 1949, Adaptation and selection, pp. 365-386 in: *Genetics, Paleontology, and Evolution*, G. L. Jepson, E. Mayr, G. G. Simpson, eds., Princeton University Press, xiv, 474 pp.

———, 1960, Physiological genetics, ecology of populations, and natural selection, pp. 429-475 in: *Evolution after Darwin*, vol. 1, Sol Tax, ed., University of Chicago Press, viii, 629 pp.

WYNNE-EDWARDS, V. C., 1962, *Animal Dispersion in Relation to Social Behaviour*, Edinburgh & London, Oliver & Boyd, xi, 653 pp.

Index

acorn storage, by woodpeckers, 201
adaptation, analogy with artifice, 10, 253; as approximation to optimum, zero, or infinity, 262-63; biotic, *see* biotic adaptation; criteria of, 4, 8-13, 17-19, 122-23, 145-46, 209-12, 252-61; to cyclic changes, 31, 71; vs. fortuitous effects, 8, 16-19, 139, 144-46, 209-12, 214, 247-48; functionally arbitrary limitations on, 263-64; general vs. special, 49; to geological future, *see* teleology; harmful consequences, 17-19, 27-28, 139, 211-12, 214-15; hierarchical organization of, 263, 266-69; intuitive recognition of, 9, 258-62; statistical consequences of, 17-19, 210-19. See *also* biotic adaptation, organic adaptation, organization, function and related terms.
adaptive radiation, 30-31
adequacy, irrelevance in evolution, 28-31
adult, as goal of development, 43-44
agonistic behavior, prevalence of, 193-94

aircraft engines, progress in, 50
alarm substance, 216
Allee, W. C., 18, 92, 105, 189, 194, 209, 218, 244, 247, 268
Allison, A. C., 60
Altman, S. A., 95, 205, 220
Amadon, D., 246, 254
amphibians, dependence of metamorphosis on diet, 67; imperfections of adaptation, 53; larval adaptation, 45; nuclear-cytoplasmic interaction in development, 61-62, 64; regeneration in, 85; reversed roles of sexes, 186; vocalization, 232
Andersen, F. S., 151
Anderson, E., 144, 145
Andrew, L. E., 64
angiosperms, replacement of gymnosperms, 51-52
annual fishes, 175
antibiotics, bacterial resistance to, 267-69
ants, population structure, 146. See *also* social insects
aphids, 132, 156
apogonid fishes, 180
apple, 6, 9
Arapaima, 180
archaic Recent fishes, 51
arctic ecology, 248-50
arctic fox, 249

INDEX

argument from design, 6
Aristotle, 258
Ascaris, 256
ascidian, 53
atomic theory, analogy with natural selection, 272-73
ATP, universality, 134
Auerbach, C., 141
autocatalysis, 5, 135
autotrophs, origin of, 248

baboons, 220
bacteria, genetic recombination among, 133, 137; quantity of DNA, 39; resistance to antibiotics, 267-68
Bardach, J. E., 165
Barker, J. S. F., 105
barnacles, cost of reproduction, 172; sessile condition of, 127
Barnes, H., 172
bass, 51, 180
Bateman, A. J., 185
bears, as predators, 16-17; evolutionary rate, 126
beechnuts, mortality rate, 164
Betta, 180
bighorn sheep, 219
biota, arctic vs. tropical, 248-50
biotic adaptation, aesthetic appeal of concept, 232-35; at community level, 246-50; criteria of, 103, 122-24, 257-58; doubts of existence, 112, 251-52; nature of, 96-98, 103-105; necessarily limited occurrence, 113; as non-parsimonious concept, 122-24; and progress, 106; and regulation of population density, 107, 234-46; relation to group selection, 96-98; mutation as, 139, 141; in penguin reproduction, 190-91; role of dominance in, 143; role of heterozygosity, 144; role of sex ratio in, 147-49; and social insects, 197-201. See also various examples
biotic evolution, and chance events, 121; and extinction, 119-22; vs. organic evolution, 96-101
Birch, L. C., 32
birds, absence of viviparity, 170; age of maturity, 90-91; clutch size, 161-62, 164-65; colonial nesting, 189; dominance hierarchies, 95; flight displays, 206; flocking behavior, 217; gallinaceous, 164-65; ground-living, 91, 181; life cycle, 90-91; low reproductive rates in, 181; misplaced reproductive functions, 204; mortality rates, 90-91, 172; nesting places, 91; optimum reproductive effort, 181; polyandrous, 186; rates of development, 90-91; reversed roles of sexes, 186; sexual dimorphism, 181; social relations among, 238

INDEX

bithorax, 72-77
Blood, D. A., 219
bluebirds, nesting requirements, 238
bluefish, 48
Blum, H. F., 38, 260
Bodmer, W. F., 154
Bonellia, 154
Bonner, J. T., 225
Bormann, F. H., 223
Borradaile, L. A., 154
bowfin, 51, 180
Boyden, A. A., 126, 128, 129, 138
Braestrup, F. W., 246
Breder, C. M., 213-16
Brereton, J. L. G., 105, 107, 232, 234, 239
Brock, V. E., 215
Brown, W. L., 49, 105
Budd, G. M., 191
Buddha, 255
budding, 126
Bullis, H. R., 215
Burkholder, P. R., 223
Burnet, F. M., 222, 230
Burtt, E. A., 255
Buzzati-Traverso, A. A., 141

caddis fly, 86-87
Cagle, F. R., 204
Cambrian, 35, 38, 42, 54, 100
canalization, 73, 75
cannibalism, 236
caribou, 219
Carlisle, D. B., 230
Carson, H. L., 104
Catcheside, D. G., 140
catfish, 165, 180
Catlin, W., 271

cause and effect, 8-9, 31, 76-78. See also teleology
centrachid fishes, 180
cercaria, 46
cetaceans, schooling behavior, 217; social phenomena, 95-96
chemical conditioning of medium, 209-10
chromosome inversions, 64
chromosome number, significance of, 132-33
cichlid fishes, 166, 180
cladocerans, 156. See also *Daphnia*
Clarke, C. A., 25
Clarke, G. L., 218
clones, mutations in, 125, selection of, 23-24
coadaptation, 59
coconuts, mortality rate, 164
cod, 165
Cole, L. C., 66, 159
colonial reproduction, 187-91
Comfort, A., 225
commensalism, 246-47.
community, organization of, 18-19, 246-50
community stabilization, 249-50
competition, ecologic vs. reproductive, 32-33; relation to natural selection, 31-33
continuous variation, vs. particulate gene, 61
copepods, fecundity, 167
Correns, C., 155
cost of reproduction, and demographic environment, 171-82; optimization of,

293

165, 171-82; related to fitness, 26; related to size and age, 182
Cott, H. B., 254
crèche system, among penguins, 190-91
Crisp, D. J., 172
criteria of adaptation, *see* adaptation
Crosby, J. L., 143
crowding, psychological damage from, 243-44
Crowe, L. K., 154
Cullen, E., 190
currency of offspring, 195, 200
cypress knees, 10
cytoplasmic influences on development, 61-64

Dalton, 272-73
Daphnia, experimental populations, 237-38; parthenogenesis of, 126-27, 132
Darling, J. F., 189
Darling-effect, 189, 268
Darlington, C. D., 7, 64, 128, 140-41
darters, 179
Darwin, C. R., 3-4, 32, 83, 94, 100, 231, 270-71
deer, depletion of browse, 17-18; herding of, 206-207; reaction to predators, 16-17, 206
degenerative evolution, 77-83, 85-86, 265-66
Democritus, 272
demographic environment, and evolution of senescence, 225-28; nature of, 66, 176, 251; and rate of development, 87-91; and selection of reproductive effort, 172-82, 245-46
density as an evolutionary factor, 29
density-governed reactions, 29
determinate growth, in fishes, 177
determinism in evolution, 21-22, 49
development rate, and population density, 235-36; selection of, 87-91
Devonian fish, complexity of, 42-43
Devonian vertebrates, numbers of surviving lineages, 120
diatom, 264-65
Dickson, G. C., 25
dietary requirements, evolution of, 265-66
Dijkgraaf, V. S., 11
dinoflagellates, 229
dinosaurs, extinction of, 121-22
dioecy vs. monecy, significance of, 147
disease, transmission of, 211-12
DNA, of amphibians, 39; biochemical instructions in, 40-42; of birds, 39; efficiency of, 41-42; as evolved adaptation, 133-35; of lungfish, 39; of mammal, 39; morphogenetic instructions in, 40-42, 46-47; optimum information con-

INDEX

tent, 38; origin of, 36, 133-35; of protists, 39, 42; quantity of, 36-42; redundancy, 36, 39, 41-42
Dobzhansky, T., 14, 59, 64, 142
dominance, evolution of, 141-44
dominance hierarchies, 95, 218
Dougherty, E. C., 128-29, 137-38
drift, see genetic drift
Drosophila, adaptation to diurnal cycle, 269; bithorax phenotype, 72-74; mutation rate, 141; relative fitness of populations, 104; sex ratio, 151-52
Dunbar, M. J., 107, 248-50
Dunn, L. C., 27, 117

Earth, uniqueness of, 121
earthworms, soil modification, 18-19; somites, 194-95
ecological environment, 57-58, 66-71
ecological necessity, not an evolutionary factor, 28-30
ecological niche, 22, 121, 264-65
economy of information, 41-42, 82-86
ecosystem, arctic vs. tropical, 248-50; efficiency of, 248; organization of, 18-19, 247-50
Edwards, A. W. F., 154
egg, as soma of zygote, 61-62

egg size, in fishes, 164, 167; and mortality rates, 164
eggs, selection for size and number, 161-64, 167-68
Ehrensvärd, G., 135, 136
Ehrman, L., 145
elasmobranch, egg size, 163
elephants, mutation rates, 141; rate of evolution, 226
Elton, C., 244
embiotocid fishes, viviparity, 180
Emerson, A. E., 28, 65, 159, 164, 225, 232, 239, 266
empiricism and theory in science, 20
empiricists vs. nativist, 83
environment, see demographic environment, ecological environment, genetic environment, social environment, somatic environment
enzymes, origin of, 135-36
Eocene, 98
epigenetic theory of development, 62, 64
Epling, C., 271
errors in plan of human body, 264
Essig, E. O., 186
ether shock, 72-75
euphausids, predation by whales, 87
evolution, degenerative, 77-83, 85-86, 265-66; deterministic stages, 134-36; reversal of, 37-38; stochastic phases, 136
evolutionary data, importance to understanding adaptation, 263-64

INDEX

evolutionary determinism, *see* determinism in evolution

evolutionary plasticity, chapter on genetic factors in, 127-57; relation to teleology, 21; role of diploidy in, 141; role of dominance in, 143-44; role of mutation in, 12, 139; role of outcrossing in, 129-31; role of population structure in, 232

evolutionary rate, and length of generation, 226; through group selection, 114

evolutionary trajectories, selection of, 110-11, 114

extinction, chance vs. fitness, 115, 121-22; through competition, 30-31; of dinosaurs, 121-22; of ichthyosaurs, 51; importance in biotic evolution, 119-22; of mososaurs, 51; not a creative factor, 121-22; through overspecialization, 27-28; of plesiosaurs, 51; of Pliocene mammals, 51; rates of, 115, 122

facultative vs. fixed character, 77-83

family groups, reduced competition within, 202-203; selection among, 196

fecundity, facultative individual control of, 237; independence of parental care in fishes, 165; nutritional factors in, 163-64; related to mortality rate, 161-64; selection of, 161-67; of social insects, 163-64. *See also* cost of reproduction, eggs, and various organisms

femininity, selection for, 182-87

ferns, 126, 131

fiddler crabs, regeneration in, 83

Fiedler, K., 186

Filosa, M. F., 224

Final Cause, 258

Fink, B. D., 214

Fisher, J., 190

Fisher, R. A., 3, 20, 21, 25, 88, 141-43, 149, 152, 154, 157, 184, 202, 273

fishes, brain structure, 43; chromosome numbers, 133; cost of reproduction, 180-81; courtship, 178; criteria of evolutionary progress, 48; demographic environment, 177-78; with determinate growth, 177; Devonian, 50; drought resistant eggs, 175; egg size, 164, 166; fecundity, 165-68, 178, 182; frequency of spawning, 178; gregarious reproduction, 188; histological specialization, 47; hybridization in nature, 205; with indeterminate growth, 176-77; integumentary histology, 43; lateral line, 10, 260; oral incubation, 180; parental care, 165, 178-80; regeneration in, 85; reproduc-

296

tive functions in relation to demographic environment, 172-81; schooling, 212-17; sex dimorphism, 175, 178; sexual conflict, 178; size and cost of reproduction, 178; territoriality, 178; toxic, 229-30; viviparous, 166-67, 178. See also specific kinds
fission, 126
fitness, as currency of offspring, 158-59; of environment, 70; inclusive, 97, 194, 196, 207; of individuals, 25-26, 101-102, 158; and levels of organization, 71; of populations, see population fitness; relation to reproduction, 25-26, 195, 200
fitness and chance in relation to survival, 101-103, 158-59
fixed vs. facultative character, 82
flatfishes, 48
flatworms, group protection, 209-10; regeneration in, 84
fleetness, 16-17
flight displays, of birds, 206; of mammals, 206
fluke (trematode), 45-47
flying fish, 11
"flying" gurnard, 260
Ford, E. B., 25
foreign nuclei, in egg cytoplasm, 61-62
fossil record, 21-22, 98-100
Fowler, J. A., 62, 64

Frank, F., 244
Freedman, L. Z., 205
friendship and animosity as evolutionary factors, 93-96
frog, see amphibians
fugitive species, 155-56
function, defined, 8-9
functional adequacy, not an evolutionary factor, 28-31
functional vs. statistical organization, 210-19, 257-58

Galileo, 20
game theory, 67-70
gametophyte of fern, 126, 131
gametophyte development, determination of, 63-64
garpike, 51, 178
gene pools, selection of, see group selection
genes, defined, 24; potential immortality, 24; origin of, 133-35; stability of, 24, 138-41. See also mutation, lethal genes
general vs. special adaptation, 49
generation length and rate of evolution, 226
genetic assimilation, 4, 7, 71-83
genetic background, see genetic environment
genetic drift, 111-15, 146
genetic environment, and dominance, 142-43; early evolution of, 137-38; and mutation rate, 139-40; role in determining selection

INDEX

coefficients, 26-27, 57-61, 119
genetic information, in algae, 40-41; in *Amoeba*, 40-41; in bithorax stock, 75-76; changing interpretation during development, 62-63; for fixed and facultative adaptations, 82-83; in humans, 40-42; quantity vs. precision, 41-42
genetic mosaics, 222-25
genetic recombination, in bacteria, 133, 137; restrictions on, 24, 132-33; significance of, 131-32; in viruses, 133
genetic vs. structural homology, 65
genetic variation, unexpressed, 73-75
genetically heterogeneous somata, 222-25
genic selection, analogy with atomic theory, 272-73; coefficients of, 23-25, 56-57, 65; effectiveness, 7, 8, 24-25; expected consequences, 25-34; and favorable characters, 26-27; of fecundity, 21, 161-67; fundamental importance, 159-60; inadequacy for biotic adaptation, 116; of isoalleles, 37; of lethal genes, 24; and maximization of individual fitness, 25-26; moral and aesthetic aspects, 4, 232-34, 254-55; and morphogenesis, 56; of multiple alleles, 58; nature of, 24-26, 71, 96-101; need for theory, 103; and the origin of man, 120-21; as parsimonious concept, 108, 123-24; and rate of development, 87-91; reversal of direction, 37, 39; role of demographic parameters, 66-67, 88, 250-52; and social environment, 93-96; in social insects, 195-201; of t-alleles, 117-19; two-locus model, 59
genotype, relation to phenotype, 56
genotypic vs. phenotypic stability, 64-65
gnathostomes, 51
goal, defined, 8-9
Gotto, R. V., 167
gourami, 179
gregariousness, of cetaceans, 217; of fishes, 212-17; as fortuitous statistical effect, 257-58; of mammals, 217; of non-breeding birds, 217; as protection from predators, 213-16; of sea snakes, 217; selection for, 206-207, 212-17, 268-69; of squid, 217; of wolves, 217-18
group-related adaptations, *see* biotic adaptation
group selection, assisted by genetic drift, 111-12; of body size, 110; chapter on, 92-124; coefficients of, 114, 122; in man, 14-15; nature of, 96-101, 109-10; need for theory, 4, 8, 103, 122; as non-parsimonious

concept, 108; quantitative inadequacy of, 110-16; of t-alleles, 117-19
guano, 13
Guhl, A. M., 218
gymnosperms, replacement by angiosperms, 51-52

Haartman, L. von, 91
haddock, 182
Hagan, H. R., 171
Haldane, J. B. S., 3, 20, 27-28, 32, 92, 142-43, 187, 195, 273
halibut, 165, 178, 182
Hall, E. R., 218
Hall, K. R. L., 220
Halstead, B. W., 229-30
Hamilton, W. D., 26, 97, 197, 199, 252
Harper, J. L., 238
helpers, among birds, 207-208
hens, social organization, 218
hermit crabs, regeneration in, 86
heterosis, 130, 144
heterozygosity, as biotic adaptation, 144; decrease in stable environment, 38
histological specialization and evolutionary progress, 47
Hodder, V. M., 182
homosexuality, prevalence of, 204-205
horses, Eocene, 98-99; fitness, 101-103; size changes in Tertiary, 98-100, 110
Hubbs, Carl L., 205
Hubbs, Clark, 167
huddling, of mice, 212

human mind, function of, 14-16
Huxley, J. S., 21-22, 49, 86, 129
hybrid inviability or sterility, 145
hybridization, as example of reproductive malfunction, 205; selection for avoidance of, 145-46
hydra, 127

immune reaction, 222
immunological evidence of genetic homogeneity, 62
inclusive fitness, 97, 194, 196, 207
indeterminate growth in fishes, 176
individuality, mechanisms for maintenance of, 222; origin of, 135-38
Inger, R. F., 231
inquilinism, 266-67
insect pollination, 27
insects, reversed roles of sexes, 186; social, see social insects; viviparous, 170-71
internal fertilization, as preadaptation for viviparity, 169-70
introgressive hybridization, 144-46
invertebrates, marine populations, 113; somatic fusion in, 222; toxic, 229-30
isoalleles, selection of, 37
Ives, P. T., 141

jet vs. propeller driven aircraft, 50

INDEX

Kelson, K. R., 218
Kendeigh, S. C., 186, 190
Kepler, 20
Kimura, M., 35-42, 81, 105, 133, 141, 143
kin selection, 26, 195-96. See also group selection
Klauber, L. M., 231
Knight-Jones, E. W., 222
Koford, C. B., 205
Kojima, K., 60
Kosswig, C., 266

Lack, D., 21, 162, 164, 170, 190, 220, 240, 242, 268
lactation, selection of, 159, 187-89
Lagler, K. F., 165
lamprey, reproductive effort of, 175; successfulness of, 51
larvae of marine invertebrates, 67
lemming, 244, 261
Leopold, A. C., 227
Lerner, I. M., 144
lethal genes, selection of, 117-19
Levene, H., 59
Levitan, M., 133
Lewis, D., 154-55
Lewontin, R. C., 27, 38, 60, 92, 105, 106, 117, 144
lichens, 246
Lidicker, W. Z., 240
life, criteria and definition, 5; origin of, 5, 134-36
life cycles, evolution of, 44-45, 70-71, 87-91, 131-32, 137-38; of ferns, 131; of higher plants, 131; of parasites, 132; relative complexity, 43-47; semelparous, 174-75
linkage, significance of, 132-33
Liosner, L. D., 84
liver, regeneration of, 86-87
liver fluke, 45-47
lizard tail, regeneration of, 83
Lloyd, M., 31
Lydeckker, R., 218

McClintock, B., 140
Makino, S., 133
malaria, resistance to, 60-61
Malthusian parameter as measure of fitness, 105
mammals, adaptive radiation of, 122; arctic, 267; brain structure, 43; flight display, 206; gregarious, 188-89, 206, 217-18; histological specialization, 47; homosexuality, 205; huddling, 212; integumentary histology, 43; litter size, 163; mutual aid among, 95-96, 218-19; selection for lactation, 159, 187-88; secondary sex differences, 183-84; viviparity, 170
mammary glands, contribution to fitness, 159, 187-89
man, adaptation to solar radiation, 82, 260; mental faculties, 14-16; origin of, 14-16, 120-21; primitive social environment, 93-96; racial differences, 15, 82; rate of evolution, 226; regenera-

tion in, 84-85; special adaptations, 121; systematic position, 49; thermoregulation, 76-77
marginal habitats, retreat of populations to, 30-31
marine invertebrates, larvae of, 90
marsupials, replacement by placental mammals, 51-52
masculinity, selection for, 182-87
Mather, K., 32
Mayr, E., 59
means, defined, 8-9
mechanism, defined, 8-9
Medawar, P. B., 158-59, 226
meiosis, origin of, 128; role in sexual reproduction, 126
meiotic drive, 27, 117-19
Melandrium, 154-55
Mendelian population, definition of, 125
Mesohippus, 47-48
mice, embryonic lethality, 27; huddling behavior, 212; t-alleles, 27, 117-19
Michie, D., 62
Miller, R. R., 165
Milne, A., 32
mimicry, 202
miracidium, 46
Mirsky, A. E., 39
misplaced reproductive functions, 194, 203-208
Modglin, F. R., 230
Montagu, M. F. A., 14, 116
morphogenesis, origin of, 135-38; role of environment in, 66-70, 72-75
Morris, D., 205

mortality rate, and selection for rate of development, 87-91; sex differences in, 150-51
mososaurs, extinction, 51
Mottram, J. C., 181
Moyse, J., 222
Muller, H. J., 9, 22
Murie, O. J., 219
Murphy, R. C., 190
mushrooms, toxic, 229-30
musk oxen, 218-19
mutation, as biotic adaptation, 12, 138-41; and direction of evolution, 100; evolutionary role, 12, 75, 138-41, 145; randomness, 37, 100-101
mutation rates, of *Drosophila*, 141; of elephants, 141; in haploid organisms, 140-41; and length of generation, 140; optimum, 139; and rate of evolution, 38, 139; relation to selection coefficients, 24-25, 57, 109-10; selection of, 138-41
mutator genes, 139-41
mutualism, evolution of, 246-47
Myers, G. S., 175

nativists vs. empiricists, 83
natural selection, and adequacy of adaptation, 21; among children, 15-16; of clones, 23-24, 109; conditions for effectiveness, 22-25; and ecological necessity, 21; and extinction,

27-28; of genotypes, 109; history of concept, 3-4; importance of phenotype, 23-26; ineffectual use of theory, 20-21, 270-71; limitations of process, 21-33; and maintenance of adaptation, 54; modern opposition to, 3-4, 7; nature of process, 22-34; and optics of eye, 6; and origin of life, 5; and physical laws, 7; related to reproduction, 26; of somata, 109. *See also* genic selection, group selection

Nature, Buddhist view of, 255; as guide in ethics and morals, 254-55; Judeo-Christian view, 255

necessity, irrelevance in evolution, 28-32

Needham, A. E., 84

Neotropical fauna, replacement by Holarctic, 52

nettles, 230

Newtonian synthesis in biology, 20

niche, *see* ecological niche

niche selection, 69-70

Nicholson, J. A., 29

Noble, G. K., 186

Norman, J. R., 228

O'Donald, P., 184

Odum, H. T., 105

optimum reproductive effort, theory of, 175-77

organic adaptations, within individual somata, 221-22; nature of, 96-98, 103-104; as parsimonious concept, 108. *See also* specific phenomena and organisms, biotic adaptation

organic evolution, vs. biotic evolution, 96-101

organic soup, 5, 134

organization, functional vs. statistical, 210-19, 257-58

orthogenesis, 21-22, 49

orthoselection, 111

ostrich, 91

owls, 30

Paleozoic animals, complexity of, 42

Paley, W., 259

Panama land-bridge, 52

parasite life cycles, 132

parental care, selection of, 159-61, 187-89

parent-offspring recognition, 188-91

Park, T., 31

parsimony in explaining adaptation, 4-5, 11-13, 18-19, 123-24, 261-62

parthenogenesis, asexual nature of, 126-27; effects on selection for sex ratio, 155-56

particulate gene, vs. continuous variation, 61

Patel, B., 172

Pavlovsky, O., 59

pearlfish, 266-67

penguins, gregarious reproduction, 190-91

Penny, R. L., 191

phalaropes, 186

phenotype, relation to geno-

type, 56; role in natural selection, 23-26
phenotypic vs. genotypic stability, 64-65
photoperiodicity, 67, 269
photosynthesis, 248
physiological necessity, irrelevance in evolution, 28-30
pike, 48
Pimentel, D., 105, 107
pipefish, 185-86
Pittendrigh, C. S., 9, 258, 260, 264, 269
placental mammals, replacement of marsupials, 51-52
plankton, oceanic, 238
plants, criteria of progress, 48, 52-53; grafting experiments, 222; hybridization in, 145-46; life cycles, 131; outcrossing, 129; population size, 238; senescence, 227-28; sex determination, 154-55; somatic fusion, 222-25; toxic, 229-30
plasmagenes, 62, 136
plasticity, evolutionary, see evolutionary plasticity
plesiosaurs, extinction, 51
Pliocene, 22, 98
Pliohippus, 47-48
poisonous flesh, as biotic adaptation, 228-31
polar bear, 28-29
pollination by insects, 27
population density, adaptive regulation of, 105, 107, 148-49, 152, 234-46; and cannibalism, 236; influence of improved organic adaptation, 29; level vs. precision of regulation, 239-40; nutritional control of, 235-37; in plants, 238; and rate of development, 235-36; and rate of reproduction, 236
population fitness, and extinction, 27-28, 49-50, 106, 119-20; and numerical stability, 105-107, 232-46; vs. population success, 103-104; in relation to size, 104-105; and sex ratio, 147-49; and versatility, 105-106
population size, see population density
populations, experimental, 235-38; misinterpretation of concept, 251; protozoan, 235
porpoises, mutual aid among, 95-96; vocalization, 10
preadaptation, 30, 169-70
pre-Cambrian, 36
precipices, instinctive fear of, 83
predation, in experimental populations, 237-38
pregnancy, cost to female mammal, 182
prephyletic evolution, 135-37
Primates, classification, 48-49; social structure, 95
primitive organisms, continued success of, 51, 53
Proconsul, 48
progress, as accumulation of information, 35-42, 46-47;

in aircraft, 50; Darwin's concept of, 100-101; as development of biotic adaptation, 106; in evolution, 4, 21-22, 34-55; as evolution in an arbitrary direction, 47-49; and gene substitution, 34-35; as histological specialization, 47; as improvement of adaptation, 49-54; as increasing complexity, 42-47; multiplicity of meanings, 35; origin of concept, 35; and precision of adaptation, 34, 49-54; vs. substitution of adaptations, 53-54

promiscuity, selection for, 185-87

Protopterus, 180

pseudo-pregnancy, 205

purpose, defined, 8-9

Rand, A. L., 217

rat, liver regeneration, 86-87

rates, *see* evolutionary rate, mortality rate, mutation rate, etc.

rattlesnake, 14, 231

redia, 46

regeneration, 83-87

repellents, integumentary, 228-31

reproduction, adapted to demographic environment, 245-46; adapted to social environment, 242-43; as biotic adaptation, 159-61, 165; cost of, *see* cost of reproduction; facultative adjustment of, 173-74, 237, 242-43; gregarious, 187-91; nutritional control of, 235-37; in relation to population density, 236; seasonal, 172-73, 249-50

reproductive effort, *see* cost of reproduction

reproductive functions, imperfection of control, 203-208; misplaced, 203-208

reptile, thermoregulation, 76-77

response to stimulus vs. environmental interference, 75-78

rhesus monkey, 95

Rhizodopsis, 43

Rich, W. H., 214

Richdale, L. E., 191, 220

Riffenburgh, R. H., 215

Ris, H., 39

Ritter, W. E., 190, 201

Roe, A., 205

root grafts, 223

rotifer, 47

roundworms, regeneration and wound healing, 47, 84

rubber, artificial, 122

Russell, E. S., 7, 19, 86

salmonid fishes, 51, 174-75

Salt, G., 202

Schmidt, K. P., 231

schooling, of fishes, 212-17; other animals, 217

scorpion, 230

sea cucumber, 266-67

sea snakes, gregariousness, 217

seahorse, 185-86

seals, 188-89

INDEX

seasonal males, 156
seasonal reproduction, evolution of, 249-50
secondary adaptations, 265-69
secondary needs, 265-69
segregation distortion, selection of, 117-19
selection, of evolutionary trajectories, 110-11; genic, *see* genic selection; interdemic and intrademic, *see* group selection, genic selection, kin selection
selection coefficients, *see* genic selection, group selection
senescence, 27, 225-28
serial homology, 87
sex determination, environmental, 154-55; genetic, 146-54; in plants, 154-55
sex ratio, evolution of, 21, 146-57, 272; role in population regulation, 148-49, 152; selection for offspring, 150-53
sexes, differing reproductive roles, 182-87; reversed roles of, 185-87
sexual vs. asexual reproduction, genic selection of, 131-32
sexual conflict, wastefulness of, 27-28; in fishes, 174-75
sexual reproduction, association with environmental change, 129-32, 137, 156; definition of, 125-27; as organic adaptation, 130-31; origin of, 133-38; in plants, 126

sexual selection, 183-85
sharks, continued survival, 51; feeding on fish school, 215; rectal gland, 10; viviparity, 180
Shaw, R. F., 154
sheep, liver fluke, 45-47
Sheppard, P. M., 25, 37
sickle cell anemia, 60-61
Simpson, G. G., 9, 21, 111-13, 122
Singer, R., 14
siphonophore, 194-95
Skutch, A. F., 207
slavery, in ants, 28
sleep, as secondary need, 268
Slijper, E. J., 96
slime molds, 223-25
Slobodkin, L. B., 92, 155, 237, 268
smelt, 113
Smith, J. L. B., 230
Smith, J. M., 242
snowshoe hare, 249
Snyder, R. L., 232, 239
social Darwinism, 255
social donors, 195
social environment, 66, 242-43, 251, 268-69
social hymenoptera, genetics of, 200
social insects, absence of viviparity, 171; analogy with human organizations, 256; caste determination, 67; evolution of social organization, 197-201; fecundity, 33-34, 163-64; with multiple queens, 199-201, 146; stings of, 230-31

INDEX

social needs, secondary status of, 268-69
Socrates, 23
somata, genetically heterogeneous, 222-25
somatic environment, 57-58, 61-67
somatic nuclei, genetic competence of, 62-63
Sommerhoff, G., 9, 260
sonar, of bats, 30
special vs. general adaptation, 49
specialization vs. evolutionary advance, 48-50
specialization and extinction, 27-28
species, analogy with human organizations, 253-54; nature and significance of, 252
spider monkey, habitat selection, 69
Spieth, H. T., 145
sponge, 53, 127
sporocyst, 46
sporophyte, of fern, 126, 131
sporophyte development, determination of, 63-64
squid, DNA of, 39; schooling behavior, 217
Stalker, H. D., 156
Stebbins, G. L., 129
sterility, 27, 117-19, 197-99
sticklebacks, egg masses, 179; homosexuality, 205
stimulus, unreliable as guide to function, 269
stingray, 230
stolon, 126
stonefish, 230
structural vs. genetic homology, 65
sturgeon, 178
sunfish, 165
Suomalainen, E., 132
superoptimal stimuli, 262-63
survival, roles of fitness and chance, 101-103. *See also* extinction
survival by specialization, 30-31
Svärdson, G., 167
swordfish, 214
symbiosis, 246-47
syngnathid fishes, 185-86
systematics as aid to understanding adaptation, 270-71

t-alleles, 27, 117-19
tapeworm, 161
teleology, 7, 21, 128, 258
teleonomy, 258-69
termites, intestinal symbionts, 246. *See also* social insects
tern, 189
territoriality, agonistic nature of, 193-94; in birds, 240-43; function of, 240-43
territory, non-feeding, 242
theoretical biology, immaturity of, 88
theory and empiricism in science, 20
thermoregulation, man vs. reptile, 76-77
Thoday, J. M., 49-50, 106
Thompson, D. Q., 244
thread herring, 215
tinamou, 186
Tinbergen, N., 242, 263

INDEX

toxins, repellent function of, 228-31
trajectories, evolutionary, 110-11, 114
transduction, 137
tree shrews, 49
trematode, 45-47
Tribolium, 31
tuna, 179
turtles, juvenile sexuality, 204; marine, 261

Underwood, G., 81
urodeles, 67

Vendrelly, R., 39
venoms, defensive, 230-31; offensive, 230
vertebrate eye, functional design of, 6, 161, 259
vertebrates, Devonian, 42-43, 120
viruses, genetic recombination among, 133; as unspecialized genes, 136-37
viviparity, in fishes, 166-67, 178; immunological obstacles, 170; in insects, 170-71; mammalian, 170; selection for, 168-71
Vorontsova, M. A., 84

Waddington, C. H., 7, 39, 49, 71-83
Warburton, F. E., 81
Wardlaw, C. W., 63
warning signals, chemical, 216; visual, 206-207
wasps, parasitic, 202-203
weaverfish, 230
Weismann, A., 3, 225
Wellensiek, U., 177
whale, 87, 264-66
White, M. J. D., 64
Williams, G. C., 61, 165, 179, 195, 196, 216, 226
Wilson, E. O., 146
woodpecker, 190-91, 201-202
Wright, S., 3, 20, 26, 58, 92, 111-13, 122, 142, 143, 196, 232, 273
Wynne-Edwards, V. C., 91, 92, 96, 105, 107, 113, 181, 187, 194, 218, 232, 239, 243-46

yeast, 13

Princeton Science Library

Edwin Abbott Abbott	Flatland: A Romance in Many Dimensions *With a new introduction by Thomas Banchoff*
Philip Ball	Designing the Molecular World: Chemistry at the Frontier
Friedrich G. Barth	Insects and Flowers: The Biology of a Partnership *Updated by the author*
Marston Bates	The Nature of Natural History *With a new introduction by Henry Horn*
John Bonner	The Evolution of Culture in Animals
A. J. Cain	Animal Species and Their Evolution *With a new afterword by the author*
Jean-Pierre Changeux	Neuronal Man: The Biology of Mind *With a new preface by Vernon B. Mountcastle*
Paul Colinvaux	Why Big Fierce Animals Are Rare
Peter J. Collings	Liquid Crystals: Nature's Delicate Phase of Matter
Florin Diacu and Philip Holmes	Celestial Encounters: The Origins of Chaos and Stability
Pierre Duhem	The Aim and Structure of Physical Theory *With a new introduction by Jules Vuillemin*
Manfred Eigen & Ruthild Winkler	Laws of the Game: How the Principles of Nature Govern Chance
Albert Einstein	The Meaning of Relativity *Fifth edition*
Niles Eldredge	The Miner's Canary: Unraveling the Mysteries of Extinction
Niles Eldredge	Time Frames: The Evolution of Punctuated Equilibria
Claus Emmeche	The Garden in the Machine: The Emerging Science of Artificial Life
Richard P. Feynman	QED: The Strange Theory of Light

Solomon W. Golomb	Polyominoes: Puzzles, Patterns, Problems, and Packings *Revised and expanded second edition*
J. E. Gordon	The New Science of Strong Materials, or Why You Don't Fall through the Floor
Peter R. Grant	Ecology and Evolution of Darwin's Finches *With a new foreword by Johathan Weiner* *With a new preface and afterword*
Richard L. Gregory	Eye and Brain: The Psychology of Seeing *Revised, with a new introduction by the author*
Jacques Hadamard	The Mathematician's Mind: The Psychology of Invention in the Mathematical Field *With a new preface by P. N. Johnson-Laird*
J.B.S. Haldane	The Causes of Evolution *With a new preface and afterword by Egbert G. Leigh*
Werner Heisenberg	Encounters with Einstein, and Other Essays on People, Places, and Particles
François Jacob	The Logic of Life: A History of Heredity
Robert Kippenhahn	100 Billion Suns: The Birth, Life, and Death of the Stars *With a new afterword by the author*
Hans Lauwerier	Fractals: Endlessly Repeated Geometrical Figures
Laurence A. Marschall	The Supernova Story *With a new preface and epilogue by the author*
Helmut Mayr	A Guide to Fossils
John Napier	Hands *Revised by Russell H. Tuttle*
J. Robert Oppenheimer	Atom and Void: Essays on Science and Community *With a preface by Freeman J. Dyson*
John Polkinghorne	The Quantum World
G. Polya	How to Solve It: A New Aspect of Mathematical Method

Hans Rademacher & Otto Toeplitz	The Enjoyment of Math
Hazel Rossotti	Colour, or Why the World Isn't Grey
Rudy Rucker	Infinity and the Mind: The Science and Philosophy of the Infinite *With a new preface by the author*
David Ruelle	Chance and Chaos
Henry Stommel	A View of the Sea: A Discussion between a Chief Engineer and an Oceanographer about the Machinery of the Ocean Circulation
Geerat J. Vermeij	A Natural History of Shells
Hermann Weyl	Symmetry
George C. Williams	Adaptation and Natural Selection *With a new preface by the author*
A. Zee	Fearful Symmetry: The Search for Beauty in Modern Physics
J. B. Zirker	Total Eclipses of the Sun *Expanded edition*

Printed and bound by CPI Group (UK) Ltd, Croydon, CR0 4YY
09/06/2025
14685663-0001